opposing viewpoints® SOURCES

nuclear arms

1990 annual

David L. Bender, *Publisher*
Bruno Leone, *Executive Editor*
Bonnie Szumski, *Senior Editor*
Janelle Rohr, *Senior Editor*
William Dudley, *Editor*
Robert Anderson, *Editor*
Karin Swisher, *Editor*
Lisa Orr, *Editor*
Tara P. Deal, *Editor*
Carol Wekesser, *Assistant Editor*

greenhaven press, inc.

PO Box 289009
San Diego, CA 92128-9009

contents

Editor's Note

Opposing Viewpoints SOURCES provide a wealth of opinions on important issues of the day. The annual supplements focus on the topics that continue to generate debate. Readers will find that *Opposing Viewpoints SOURCES* become exciting barometers of today's controversies. This is achieved in three ways. First, by expanding previous chapter topics. Second, by adding new materials which are timeless in nature. And third, by adding recent topical issues not dealt with in previous volumes or annuals.

Viewpoints

The Arms Race

1. **The Arms Race: An Overview** *by Robert A. Strong*   1
   Since the end of World War II, the U.S. and the Soviet Union have spent increasing amounts of money to build more and more nuclear weapons. Whether and how this arms race will end is the subject of much debate.

2. **The Arms Race Is Over** *by Kosta Tsipis*   7
   Now that the Soviet Union has ended the Cold War and is asking for dramatic cuts in nuclear arms, it is time for both superpowers to declare the end of the arms race and focus on the problems of the environment and the economy.

3. **The Arms Race Is Not Over** *by Sidney D. Drell*   11
   Even if the superpowers destroyed their nuclear weapons, other nations would undoubtedly keep their own stockpiles. Rather than abolish their weapons, the superpowers should continue to use deterrence as their primary nuclear policy.

4. **U.S. Nuclear Policy Is Adapting to Soviet Reforms** *by James A. Baker III*   15
   Changes in the Soviet Union have prompted the Bush administration to reevaluate nuclear strategy. The U.S. can now reduce the overall number of nuclear weapons it has, but must maintain a stockpile to deter other nations from using nuclear weapons.

5. **U.S. Nuclear Policy Is Not Adapting to Soviet Reforms** *by Fred Charles Iklé*   21
   Political changes in the Soviet Union require that the U.S. make decisive changes in its defense policy. The Pentagon's insistence on continuing outdated deterrence policies only hurts the U.S.

Economics of the Arms Race

6. **The U.S. Should Cut Defense Spending** *by Gene R. La Rocque*   27
   The U.S. has more than enough weapons to defend itself. There is no need to spend large sums of money to build more nuclear warheads.

7. **The U.S. Should Not Cut Defense Spending** *by William J. Crowe Jr.*   29
   The Soviet Union remains a military threat to the U.S. despite its new policy of perestroika. It would be shortsighted to cut U.S. defenses at this time.

8. **Cutting Defense Spending Will Help the U.S. Economy** *by Karen Pennar & Michael J. Mandel*   31
   Cutting the defense budget will lower the budget deficit, inflation, and interest rates thus leading to a healthier, faster-growing economy.

9. **Cutting Defense Spending Will Not Affect the U.S. Economy** *by Murray Weidenbaum*   35
   Defense spending has not been shown to have a significant impact on America's economy. Other factors, such as trade imbalances, have far more influence.

10. **Defense Spending Should Be Cut to Fund Social Programs** *by Jack Beatty*   41
    The U.S. does not need to maintain its current level of military force. The money now being spent on the military would be better spent on programs such as national health care.

11. **Defense Spending Should Not Be Cut to Fund Social Programs**   47
by Robert F. Ellsworth

> The U.S. must maintain a strong military that can react quickly if superpower relations worsen or if a new threat to world stability arises. Funding for domestic social programs should not come from cuts in the defense budget.

12. **Improved Soviet-American Relations Warrant Defense Cuts**   51
by George J. Church

> In view of the changes in Eastern Europe and the Soviet Union, defense spending should be cut by more than has been proposed. Deep cuts would save the U.S. billions of dollars and indicate to the Soviets that Americans are ready to live in peace.

13. **Improved Soviet-American Relations Do Not Warrant Defense Cuts**   55
by John Walcott

> Although Soviet-American relations have improved, there is no guarantee that this will last. A period of détente is not an adequate reason for cutting U.S. defenses.

Arms Control

14. **The Cold War's End Proves the Benefits of Arms Control**   59
by Joseph S. Nye Jr.

> The real virtue of arms control is that the negotiating process improves superpower relations. These improved relations are partially responsible for ending the Cold War.

15. **The Cold War's End Proves the Irrelevance of Arms Control**   63
by Kenneth L. Adelman

> The Cold War is ending not because of arms control negotiations but because Soviet communism is collapsing. Arms control has never been an effective way of reducing superpower tensions.

16. **The U.S. Should Negotiate a START Agreement** *by Robert S. McNamara*   69

> The U.S. should take advantage of warmer relations with the Soviet Union and negotiatiate cuts in many categories of nuclear weapons. The Strategic Arms Reduction Talks can generate an agreement that will reduce nuclear stockpiles.

17. **The U.S. Should Negotiate a Conventional Arms Agreement** *by Jay P. Kosminsky*   73

> Signing a conventional weapons treaty with the Soviet Union as soon as possible will reduce the Soviet military presence in Eastern Europe. A START treaty is less likely and will do less for U.S. national security.

18. **Negotiating Cuts in Naval Weapons Would Promote U.S. Interests**   77
by Michael L. Ross

> The U.S. Navy has traditionally been reluctant to engage in arms control. Such reluctance is no longer justified now that the Cold War has ended. Naval arms negotiations would promote peace.

19. **Negotiating Cuts in Naval Weapons Would Harm U.S. Interests** *by Carlisle A.H. Trost*   83

> The U.S. has always relied in part on naval power for its world stature. National security would be endangered if the U.S. negotiated reductions in its naval forces.

20. **Arms Control Will Lead to Peace** *by Wolfgang Altenburg*   89

> By increasing communication between the superpowers, arms control negotiations provide an atmosphere of peace and stability. The resulting arms control agreements reduce the number of nuclear weapons, yet maintain a deterrent force.

21. **Arms Control Alone Will Not Lead to Peace** *by John M. Swomley*   93

> Simply reducing the number of nuclear weapons through arms control is insufficient. The Soviet Union's efforts toward unilateral disarmament prove that eliminating nuclear weapons is the most effective path to peace.

The Stealth Bomber

22. **The Stealth Bomber Will Improve U.S. Defense** *by Gregg Easterbrook*   97

> The Stealth bomber is an ideal weapon now that the Cold War has ended. It could replace older, more dangerous systems, thereby increasing U.S. security.

23. **The Stealth Bomber Will Harm U.S. Defense** *by Art Hobson & Jeffrey Record*   101

> The Stealth bomber is too impractical and expensive. Funding the bomber will prevent the U.S. from developing other nuclear weapons more capable of defending the country.

24. **The Stealth Bomber Should Be Funded** *by William B. Scott*   105

> The Stealth bomber deserves full funding because its ability to elude Soviet radar will improve U.S. national security.

25. The Stealth Bomber Should Not Be Funded *by Michael Brower*   109
The Stealth bomber is far too expensive and destabilizing a weapon to justify its continued development.

Bibliography   B-1

Index   I-1

The Arms Race: An Overview

Robert A. Strong

At least since the time of Thucydides, it has been considered respectable to spend time examining the past and speculating about the causes of phenomena in human affairs. Only more recently has it become fashionable for scholars to call themselves students of the future. There are, no doubt, several reasons for this development. It is very likely that with the surplus of Ph.D. candidates in recent years, graduate students desperate to find dissertation topics found that there was simply not enough past to go around and turned their attention to the future. It is equally plausible that laziness has had something to do with it. There are obvious temptations for college professors to conduct their required research in areas where there are no documents to read, no archives to visit, no policy makers to interview, and no one who can prove (at least not immediately) that a particular set of conclusions is absolutely wrong. There is a third, and more serious, reason for scholarly interest in the future—the recognition that we live in dangerous times, when questions about whether the human race will be able to avert monumental man-made disasters have become issues of the highest importance.

When scholars do examine the future, whether for frivolous or serious reasons, they do so with considerable trepidation. Historians and political scientists are naturally uncomfortable with crystal balls, tea leaves, tarot cards and the other paraphernalia of professional fortune telling. They need some basis for making predictions, some sort of evidence to evaluate, some grounding in the past for what they expect in the future. One solid source for such analysis can be found in the statements made by leading figures of the nuclear age—statesmen, strategists, scientists, intellectuals, nuclear protest leaders and others—who in the past 45 years have

said or written what they think will happen to a world armed with enormously destructive weapons. Scholars can, in other words, study the history of nuclear futures. Such a study naturally raises basic questions about the nuclear age, questions that are getting harder and harder to answer. What follows is one scholar's modest effort to summarize the history of nuclear futures, organized around four fundamentally different versions of how the nuclear age will play out.

Inevitable Nuclear War?

The first and foremost of the predictions about the nuclear future is the one that proclaims the inevitability of catastrophic nuclear war. This is the prediction made most often by science fiction writers, by the public at large during periods of heightened cold war tensions, and by nuclear protest leaders. The arguments in support of this prediction are familiar. Over time, the longer we are in possession of nuclear arms, the more likely that there will be accidents, or miscalculations, or uncontrollable escalations of confrontations and proxy wars. In periods of crisis, all of these dangers are, of course, magnified. And even if some form of detente between the superpowers were to diminish the likelihood of great power nuclear war, the process of nuclear proliferation would only move the same set of problems to dangerous and unstable regions of the world. In the long run, the more nations acquiring nuclear weapons, the harder it will be to control or regulate their use, and the greater the chances of disastrous mistakes. For some observers this is, at its heart, a simple mathematical proposition. The more weapons there are, the longer they are in existence, the greater the probability of their use. The underlying political assumption behind such calculations is the proposition that political leaders and governments cannot be trusted with instruments of ultimate destruction. Given what we know about human nature and political history, this is

Robert A. Strong, "The History of Nuclear Futures," *Arms Control*, May 1989. Reprinted with permission.

a difficult proposition to dispute. History does seem to show that war is rarely what Carl von Clausewitz wanted it to be—a rational instrument of policy—and much more often what he feared it would be—an uncontrollable spasm of violence.

The prediction of probable, and very often near-term, nuclear war is the hallmark of most nuclear protest movements. It is, in fact, the engine that drives those movements. Whether one reads Bertrand Russell writing in the late 1950s and early 1960s or Jonathan Schell and E.P. Thompson writing 20 years later, the refrain remains the same. Nuclear war is both horrible to imagine and likely to occur. A typical statement from the British nuclear protest movement nearly three decades ago puts the case plainly:

> Every day and at every moment of every day a trivial accident, a failure to distinguish a meteor from a bomber, a fit of temporary insanity in one single man, may cause a nuclear world war, which, in all likelihood, will put an end to man and to all higher forms of animal life.

Except for the addition of some mention of faulty computer chips and the likelihood of nuclear winter, nothing in this statement would have to be changed in order to use it in the American freeze campaign or the European nuclear protest movements of the early 1980s.

The problem with these predictions in nuclear protest literature is that they are very often the end of what is said about the problems we face. The presumption is apparently made that if we become convinced that nuclear war is likely, and if we also become aware of the horrors it will bring, we will automatically be motivated to take radical political steps to avert the dangers we have foreseen. Jonathan Schell's popular account of the possibility of nuclear extinction, *The Fate of the Earth,* ends with a vague plea to 'reinvent politics' and 'reinvent the world'. Not much is said, or presumably needs to be said, about how these reinventions are to take place. This is the *Christmas Carol* account of the nuclear future. World leaders and the populations of Western democracies will, like Ebenezer Scrooge, be miraculously transformed after having been visited by the ghosts of Hiroshima's past, the present arms race, and the graveyard future that awaits us all in the aftermath of a nuclear holocaust.

Deterrence Forever

Though it is somewhat less popular and far less dramatic than the speculations about an approaching nuclear war, the second version of the nuclear future—perpetual deterrence—has been much more influential. It has, for the most part, been the dominant view of policy makers in the postwar world. Those who believe that deterrence—the prevention of nuclear war by the threat of nuclear retaliation—has worked well in preventing war between the United States and the Soviet Union during the years since 1945 usually express confidence that it will continue

to do so for some time to come. This widely held confidence does, however, have limits; and the advocates of nuclear deterrence are often reluctant advocates. Their reluctance comes from several sources.

"The possibility always exists that some . . . insane leader will prove that the nuclear protesters have been right all along."

A policy of nuclear deterrence raises obvious moral and practical problems since it depends on making plausible threats to carry out horrendous retributions against an enemy who fails to respect our territorial integrity and vital interests. In Winston Churchill's classic formulation of the perils that accompany deterrence he tells us that in the nuclear age, 'Safety will be the sturdy child of terror, and survival the twin brother of annihilation.' Living with terror and the threat of annihilation are the high, but perhaps unavoidable, costs that must be paid in order to sustain a policy of nuclear deterrence.

Related to the recognition that deterrence rests on shaky moral ground is the equally disturbing prospect that it may be imperfect. It might fail. The possibility always exists that some accident, or miscalculation, or insane leader will prove that the nuclear protesters have been right all along. Churchill, in the same speech quoted above, warned the British parliament that 'I must make one admission, and any admission is formidable. The deterrent does not cover the case of lunatics or dictators in the mood of Hitler when he found himself in the final dug-out. That is a blank.' Supporters of a policy of deterrence do not, as a rule, deny the allegations made by nuclear protest leaders regarding these dangers. The difference between them is their estimates of how great these risks are and whether or not they can be tolerated and managed. Proponents of nuclear deterrence, while never really enthusiastic about the prospects of living in a world where nations are armed with enormously destructive weapons, have, nevertheless, been willing to do so in order to gain the benefits—in terms of safety and survival—that come from such an arrangement and because they do not believe that any viable alternative exists. For many of its strongest proponents, nuclear deterrence is the policy we should follow in the future for the largely negative reason that no better policy seems to be available.

The preceding paragraphs discuss nuclear deterrence as if it were one thing. In fact, it is many. Every self-respecting strategist in the years after 1945 has invented some form of deterrence and written extensively about its nature and prospects. There is 'extended' deterrence, 'existential' deterrence, 'minimum' deterrence, and 'mutual' deterrence to mention only those beginning with the letters *e* and *m*. Serious scholars have explored the problems of what deterrence is, how it works, and what will be necessary to make sure that it continues to work in the future. Since no one knows for sure what goes on in the minds of our enemies, these issues are hard to resolve and many plausible statements about deterrence can legitimately be made. Some say that the Russians are deterred from invading Western Europe because they are afraid of NATO's [North Atlantic Treaty Organization] conventional capabilities, or NATO's tactical nuclear weapons, or America's strategic arsenal, or the existence of independent British and French nuclear forces, or because they were never much interested in invading Western Europe in the first place. The difficulty in providing a definitive determination about which one of these statements is the most accurate has not kept scholars and commentators quiet. Quite the opposite has occurred; we have had endless academic and public debates about NATO's nuclear strategy, and whether or not particular weapon systems, like the Pershing II, or particular arms control agreements, like the INF [intermediate-range nuclear forces] agreement, would enhance or weaken the deterrent effect that is nearly always the universal goal. The same observations could be made about debates over strategic arms and our ability to deter a Soviet nuclear attack on the United States. What is remarkable about these debates has been the relatively narrow range in which they have taken place. Critics of existing nuclear policies, whatever those policies may be, usually present themselves as the true friends of deterrence. Even advocates of limited nuclear war generally make their case by pointing out the need to deter different kinds of Soviet threats and to have additional options for our national leaders in the unlikely event that deterrence should fail.

"What if a nuclear war could really be won?"

The limited nuclear war option, though traditionally defended as an augmentation to deterrence, does occasionally slip across the line separating nuclear futures and become an independent and viable alternative to either catastrophic nuclear war or deterrence forever. Some people believe that technical improvements in nuclear weapons (making them smaller, more accurate, more easily controlled by their commanders), would make it possible successfully to employ them in a modern European or global war. What if a nuclear war could really be won? Other proponents of a technological alternative to our current nuclear morass ask equally

provocative questions. What if a meaningful defense to nuclear attack could be deployed? What if some fantastic new weapon system, whose scientific principles may as yet be unheard of, were to be discovered and developed? What if technology could get us out of the mess it has gotten us into? The third version of the nuclear future—the technological fix—is difficult to discuss without a question mark.

For almost all of the nuclear age and for almost all of the political leaders who have served in that age, the possibility of a technological fix has been little more than a possibility. It has been worth thinking about and worth spending money on, sometimes large amounts of money, but its success has always been a dubious proposition. Even when the United States was actively building anti-ballistic missile systems (ABMs) in the late 1960s and early 1970s, political leaders tended to describe their capabilities in modest terms. Robert McNamara publicly claimed that the ABM system ordered by the Johnson administration would not defend the country against a full-scale Soviet attack, but might offer some protection against small nuclear powers like China. Ronald Reagan's arguments for a star wars defense have been rather different than those used to support ABMs some twenty years ago. Reagan promised a defensive shield that would render nuclear weapons 'impotent and obsolete'. And though many of his advisers discounted the president's promise as rhetorical excess and described the strategic defense initiative in traditional terms as a way to strengthen deterrence and as an insurance policy against the remote possibility that deterrence would someday fail, it is precisely the promise of perfect defenses that gave the president's program its enormous political appeal. The president of the United States, prematurely in the view of most scientists and strategists, asked the ultimate, 'What if?' concerning our nuclear future. He asked if there might not be some technological alternative to the dangers of deterrence and the chances of catastrophic nuclear war that virtually all other postwar political leaders have reluctantly accepted. What may be most interesting about Reagan's simple question is not its scientific infeasibility or its strategic naiveté, but its remarkable political appeal and the fact that it has only rarely been asked by serious political leaders in the nuclear age.

A New World Order

The fourth and final version of the nuclear future is, like the technological fix, hardly ever advocated by prominent political leaders. It involves dramatic changes in the nature of international politics that would not only make nuclear war very unlikely, but would hopefully make war of all kinds a thing of the past. Arms control and disarmament, in the view of those who seek a new international order, can never fully solve the problem of war, just as gun control can never fully eliminate crime. Attention must be paid to root causes and to fundamental issues. While this may be an intellectually consistent and admirable point of view, enormous practical problems arise when one begins to consider how to go about producing the kinds of international change that would be needed to outlaw war. For some observers in the nuclear age, the threat of nuclear holocaust ought to be a sufficient inducement to overcome whatever roadblocks may exist on the path to world government. 'Before the atomic bomb', Robert Hutchins, the influential president of the University of Chicago, observed in 1947, 'we could take world government or leave it. We could rely on the long process of evolution to bring world community and world government hand in hand. Any such program today means another war, and another war means the end of civilization. The slogan of our faith today must be, world government is necessary, and therefore possible.' Hutchins' plea, like Schell's 25 years later for the reinvention of politics, states a goal, but does not say very much about how to get there. Those who do have plans for the creation of a world government—the world federalists, or the proponents of a greatly strengthened United Nations—are generally regarded as hopelessly naive and have no discernable political power. . . . Calls for a new international order have, for the most part, not been taken seriously in the nuclear age, despite widespread recognition of the consequences of nuclear war and the dangers of deterrence. Even Jonathan Schell, Hutchins' intellectual heir, quickly retreated from his call for a wholly new politics and, in his second book on nuclear issues, became the advocate of an unusual version of deterrence which he believes would be safer than the one we currently employ.

"Even before the first bombs were dropped on Hiroshima and Nagasaki, perceptive scientists . . . were warning . . . about the dangers of an arms race."

An even greater problem, for those who wish to see a new international order, than the problem of how to get there, is the possibility that once we have created such an order we might not like it very much. The typical American assumption is that a world government would be very much like an American government, created in a nearly bloodless revolution, with checks and balances, a central administration that protects individual rights, and a federal structure that allows for cultural diversity and local autonomy. Too many observers assume the *Star Trek* version of a new world order, where hundreds of years from now every race and nationality remains fully identifiable

and peacefully federated, where crewmen from Scotland, Russia, and Japan (and from the planet Vulcan) work side by side under the wise supervision of someone who is obviously an American. Such a future may be pleasing to our imaginations, but it may not be very realistic. History suggests that when large parts of the globe have been under the direction of a single government, that government has won its position with military power and maintained it with liberal amounts of repression. Empire, rather than federation, may offer a more likely description of a future world government; and an empire, even one that made nuclear war impossible, might have significant drawbacks in other respects. It is precisely the fear of a hostile world empire that has created one of the most lasting clichés of the nuclear age, 'Better red than dead.' If death were the certain and sole alternative to communism, and if death were to come in the form of a massive nuclear holocaust, there might, indeed, be something to be said for becoming a collaborator with the powers that could prevent such a disaster. But, as a rule, we have ridiculed the red or dead choice because other options, like living with deterrence, have been seen to exist. But how much longer will the deterrence option continue to be viable?

Debates About the Nuclear Future

All four versions of the nuclear future—the fear of nuclear war, the confidence in deterrence, the search for a technological solution, and the dream of a world government—were part of a lively debate at the outset of the nuclear age. Even before the first bombs were dropped on Hiroshima and Nagasaki, perceptive scientists like Neils Bohr and Leo Szilard were warning political decision-makers about the dangers of an arms race and a possible nuclear war. Almost as soon as the Second World War was over, strategists like Bernard Brodie were beginning to work out the logic of deterrence. And while the nuclear age was still young, there was tremendous hope that some technological or political solution could be worked out which would turn fissionable material into cheap electrical power and the United Nations into a major international institution. The open debate about the future of the nuclear age came to a close at about the time of the Soviet rejection of the Baruch Plan and the onset of the cold war. Thereafter, deterrence became the dominant American policy and the dominant vision of the future.

Deterrence's central position in the minds of policymakers was, however, always insecure. It was constantly challenged by strategists who pointed out that the balance of terror was delicate and in need of technological improvement and arms control management in order to ward off war-winning expectations by the other side. It was equally challenged by peace advocates who pointed out that deterrence might fail and that any nuclear war would be calamitous and immoral. For the most part, promises of meaningful defenses and effective world government played minor roles in the postwar debates, but we are in a period when that may be changing and when the debate about the nuclear future may become much more volatile.

Ronald Reagan came to office in 1981 as the leader of the most effective attack on deterrence from the right that we have ever seen. He and his supporters arrived in Washington talking about Soviet strategic superiority, the futility of arms control, and the possibilities of actually fighting limited and protracted nuclear wars. Within a very short period of time, these statements helped to induce the greatest assault on deterrence from the left that we have ever seen. Massive peace movements in the United States and Europe were joined by establishment forces—physicians and Catholic bishops—who took up the traditional arguments of the peace movements and gave them added weight. Doctors warned us that the medical consequences of a nuclear war would be far worse than most military strategists had envisioned because our modern complex health care system would not survive the use of even a few nuclear weapons. The Pastoral Letter of the Catholic bishops all but said that deterrence was an unacceptably immoral policy. The only concession the bishops made was for a temporary continuance of deterrent forces while a sincere search for some alternative was conducted. But what alternative?

At the beginning of his third year in office, Reagan shifted his position. He stopped talking about fighting nuclear wars and instead put forward the suggestion that a technological fix, a star wars defense, was within the grasp of modern science. Administration arguments in support of the strategic defense initiative, to a surprising extent, have adopted the criticisms of deterrence that were once the exclusive property of anti-nuclear protesters. A defensive shield is needed because deterrence may fail; there may be accidents, miscalculations, or lunatics who attempt to attack the United States. Defensive weapons are superior to offensive ones because they kill missiles and not people and rescue the American public from the horror and the guilt of living with the possibility of nuclear annihilation. The problem with Ronald Reagan's alternative is that it is widely believed by experts in the various technologies involved to be far beyond our ability to achieve and likely to stimulate an expensive competition with the Soviet Union that will produce little net change in the deterrent capabilities of either superpower.

Future of Deterrence

A unique arms control opportunity now faces the Bush administration in its negotiations with the Soviet Union. Reasonable restrictions in star wars development and deployment can probably be traded for significant reductions in the superpower nuclear arsenals. Those reductions would not take us out of the world of deterrence, but neither would the deployment of star wars systems that were only partially effective and susceptible to Soviet countermeasures. The clear message is that, for the foreseeable future, we will continue to live in the world of deterrence, with all the terror and all the danger that Winston Churchill warned us about more than 30 years ago. . . . There may be renewed international efforts to slow the process of nuclear proliferation, but in a world with many bitter and unresolved regional disputes, the increased ease with which nuclear weapons can be built will remain a powerful temptation. What then will happen when President George Bush offers the American public the traditional vision of a future in which the United States, and others, will remain armed with weapons of mass destruction and its security, such as it is, will remain dependent on a nuclear deterrent?

"Radical alternatives to the strategic status quo have received more serious attention in the last decade than at any time since the birth of the nuclear age."

Even more than in earlier periods of the nuclear age, the ideas that surround the policy of deterrence have become intellectually damaged goods. Attacked from the right, attacked from the left, set aside by a temporary dream that new technologies would rescue us from our nuclear dilemmas, deterrence has had to withstand a great deal in the 1980s. Radical alternatives to the strategic status quo have received more serious attention in the 1980s than at any time since the birth of the nuclear age. It is an open question whether the American people can continue to be persuaded to accept a vision of the nuclear future in which our safety and survival will forever depend on weapons of terror and threats of annihilation. More importantly, it is increasingly an open question whether they should be.

Robert A. Strong is the chairman of the politics department at Washington and Lee University in Lexington, Virginia.

"Continuing the arms race as a sign of U.S. resolve to match an adversary who is offering to abandon the competition doesn't make much sense."

The Arms Race Is Over

Kosta Tsipis

Arms control efforts in the United States following World War II have mainly focused on opposition to specific nuclear weapons systems and dangerous defense policies. These efforts have largely consisted of persistent advocacy for negotiated agreements with the Soviet Union to limit or reduce nuclear arsenals, and of scientific studies setting forth the technical or operational flaws of various nuclear weapons systems. And there has been some degree of success: nuclear war has been avoided; deployment of destabilizing antiballistic missile systems has been averted; but, most importantly, arms reduction agreements have become legitimate components of U.S. national security policy.

In the past 45 years, arms control activities have been determined by the environment of hostility and mistrust between the U.S. and Soviet governments, an environment created by threats to the Western democracies emanating from a truculent, militarily powerful Soviet Union. Because these threats were warlike, the U.S. response was in the same vein, albeit often overreactive. During that time, it was mainly up to the arms control community to invent and advocate weapons reduction initiatives, suggest ways to verify them, maintain working contacts with Soviet colleagues, and oppose dangerous Pentagon plans and weapons.

But now an entirely new era has emerged: the Soviet Union is renouncing the Cold War and seeking drastic reduction of nuclear and conventional military forces, while U.S. allies are clamoring for deeper cuts in nuclear weapons in Europe. The United States and the Soviet Union are conferring at several levels: negotiations to reduce nuclear arsenals and shrink conventional forces in Europe are now under way; unprecedented arrangements for on-site verification inspections in support of arms control agreements are already in place. And the nature of the threat has changed: The emerging current dangers to the United States are not warlike; rather, the looming threats are environmental degradation and economic decline.

In this new climate, it is perhaps timely for arms control advocates to reconsider the direction of our efforts. Instead of continuing the microcriticism of the Pentagon's weapons wish-list, we should focus attention on a new set of tasks. We should, for example, develop a new U.S. agenda that defines our concepts of U.S.-Soviet relations, along with proposals for dealing with the new threats, both national and global. Or, intermediately, we should oppose undesirable weapons, not piecemeal but in the context of reexamining the missions, the functions, and the role of the military in the post-Cold War era. No such examination has been undertaken since President Harry Truman ordered one in 1946 just after the war.

Changing Role of the Military

Many of the missions for which the military traditionally prepares are becoming less relevant to national security. Traditionally, the military is expected to protect the United States from external threats by guarding its borders, and to protect its economic interests by securing markets and sheltering allies across as large a section of the world as the United States can afford. To back up these missions, U.S. armed forces have stood ready to oppose aggression and favorably resolve conflict by fighting and winning wars.

It is no longer possible, however, to resolve any conflicts with the chief U.S. adversary, the Soviet Union, by fighting. To resolve conflict, war must result in an asymmetrical final outcome: winner and loser. But if two combatants use nuclear weapons, the damage on both sides will be indistinguishable. Because the winner-loser distinction is lost, nuclear wars cannot resolve conflicts.

Kosta Tsipis, "New Tasks for Arms Controllers," *Bulletin of the Atomic Scientists,* January/February 1989. Reprinted by permission of the BULLETIN OF THE ATOMIC SCIENTISTS, a magazine of science and world affairs. Copyright © 1989 by the Educational Foundation for Nuclear Science, 6042 S. Kimbark Ave., Chicago, IL 60637.

There is no prospect of victory in even a "limited" nuclear exchange, nor in conventional combat between nuclear powers. The outcome will still be symmetrical. In the first case, there will be rapid nuclear escalation, with the hope that the other side will cry "Uncle!" first, ending up with equally devastated societies. And conventional combat will lead either to a stalemated battlefield or, if one side starts winning, to escalation to full nuclear war, initiated by the losing side in order to avoid the humiliation of defeat. Again, no net winner, and no resolution of the conflict. Deterrence, the acknowledged symmetrical state of mutual vulnerability, is the de facto recognition that neither the United States nor any other nuclear power can win a war; nuclear nations can only jointly protect the peace.

No Economic Role

The economic function of the military is also vanishing. The armed forces can no longer be expected to play a mercantilistic role in a country's economy. In the past, the military could capture raw materials or markets to benefit a nation. But the most important economic resource now is trained personnel, and people cannot be forced to produce by military means. Capturing the complex and delicately balanced industrial base of a country would yield no benefits. Imagine what would happen to the industrial productivity of Japan if the Soviets occupied that country.

"Neither the United States nor any other nuclear power can win a war; nuclear nations can only jointly protect the peace."

Keynes notwithstanding, indications abound that maintaining a large military force saps a country's economic vitality. The Soviet Union is in deep economic trouble and the global economic dominance of the United States is being seriously eroded by smaller, more militarily restrained nations like West Germany and Japan. The United States spends 6.4 percent of its gross national product for defense, as against Germany's 3.1 percent and Japan's 1 percent. But the really telling numbers are the percentages of research and development funds spent by the military in those countries. For Japan the figure is 4 percent, for Germany 8 percent, while the U.S. military controls the spending of fully 75 percent of federal R&D money. As a result, America's military technology remains supreme, but its industrial preeminence is seriously declining. In the past, conflicting economic interests might have been resolved by combat, but resort to war in pursuit of economic goals is now doubly impractical, because the United States' chief economic competitors are its own allies.

The notion that U.S. military power protects U.S. allies from Soviet dominance and thus generates a favorable economic climate for the United States is outdated. As relations between other industrialized democracies and the Soviet Union improve, the U.S. military umbrella becomes progressively irrelevant to the allies' security and is thus decreasingly significant in their economic calculus toward this country.

Treasures Without Boundaries

The clearest "present danger" facing the United States today is environmental deterioration, both local and global. The "greenhouse effect," ozone depletion, acid rain, pollution of all kinds are threatening the air, the water, the very ground we live on. Unlike raw materials, these treasures are not contained within national boundaries. They cannot be captured as the spoils of victory, or secured by military force. They are indivisible. The healing and preservation of these global resources require cooperation rather than combat. Military might is irrelevant in the battle for the environment.

The Cold War is waning. Margaret Thatcher and others have gone so far as to say it is over. In this warming zone, even the *perception* of military power is devalued; continuing the arms race as a sign of U.S. resolve to match an adversary who is offering to abandon the competition doesn't make much sense.

With economic and environmental threats pressing on all sides, with many of the traditional functions of the military rapidly becoming obsolete, it is time for the arms control community to examine critically the size, structure, and functions of the military. Both nuclear and conventional forces are needed to maintain deterrence and sustain global stability. But we need to reevaluate the level of expenditure necessary to support these missions, and then devise institutional structures and procedures to redirect some of the scientific and technical resources now consumed by the military to address present dangers: economic decline and environmental degradation.

Arms control adherents have been slow to recognize that hitherto important tasks now have decreasing payoffs. Many of us are still preoccupied with efforts useful only in the Cold War environment, when arms control advocacy came mainly from "outsiders." During the Cold War, discourse among scientists and arms controllers was often the only open channel of communication between the United States and the Soviet Union; but today, expending treasure and time on improving working relations between U.S. and Soviet scientists is like trying to break down an open door. It is perhaps a measure of the cascading changes now occurring in U.S. relations with the Soviets that these efforts, while still valuable, appear far less urgent and vital than they did just a few years ago.

Should we not begin to focus on the broader issues of the proper role of the U.S. military and on the scale of scientific and technological resources the Pentagon can justifiably claim for future development of new weapons?

Focus on the Future

The arms control community has convinced the publics of East and West alike that nuclear war, in any form, cannot be won and therefore must never be fought. National governments have adopted the nuclear weapons reductions agenda proposed by the community. The economic ravages that exorbitant military expenditures have visited upon the economies of the United States and the Soviet Union are imposing a measure of restraint on the military oligarchies of the two nations. To continue with episodic opposition to specific weapons systems—be they the homeless MX or the spherically foolish B-2—while leaving the overall role of the military unexamined would be stunningly shortsighted. The focus should now be on the problems of the future rather than on the ghosts of the past.

"The focus should now be on the problems of the future rather than on the ghosts of the past."

It is perhaps time to reflect on whether arms control efforts should continue with the momentum acquired over the past four decades, or instead shift direction and tackle new, more difficult, but increasingly relevant issues. It is undoubtedly easier to hoe the familiar row but is it our best choice? Responsible activism has a continuing valuable role to play not so much in controlling arms but in contributing to the informed *triage* the nation will have to perform on the military establishment in order to meet the new threats.

Kosta Tsipis is the head of the Program in Science and Technology for International Security at the Massachusetts Institute of Technology in Cambridge, Massachusetts.

"A major change . . . in the world political order will be required before we can abandon a policy of deterrence based on strategic offensive nuclear forces."

viewpoint **3**

The Arms Race Is Not Over

Sidney D. Drell

We are debating the resolution that nuclear weapons should be abolished by the year 2000 because of the remarkable meeting at Reykjavik, Iceland, where Ronald Reagan and Mikhail Gorbachev apparently came to the brink of concluding that all nuclear-armed ballistic missiles and perhaps all nuclear weapons should be abolished by the end of the century. Those Reykjavik discussions, which made our European allies so nervous, are commonly ridiculed, but I do not share that view. I believe that Reykjavik made the subject of significantly reducing the number of nuclear weapons, not just limiting or controlling them, an issue for serious public discussion. Before that, such discussions were mostly confined to academic seminars or peace group panels.

It is good to get back to basics: Can we get rid of nuclear weapons? Are they usable or are they so destructive that civilized human beings cannot even think of "pushing the button"? Over the years we have grown too accustomed to their face. We not only accept them but we have elevated them to the fine art of battle plans and nuclear war-fighting scenarios. We need to be reminded of what President Dwight Eisenhower said in 1956: with these weapons of mass destruction we are rapidly reaching a point where no all-out war can be won—because it is no longer a battle to exhaustion and surrender, but "destruction of the enemy and suicide."

A Distant Goal

The destructive potential of nuclear weapons is so great and their murderous impact so indiscriminate that leaders from all walks of life have concluded that they must never be used. If these weapons are not usable, if their existence may threaten our very existence and leaves little if any margin for error, then

inevitably we ask, "Why not get rid of them—and the sooner the better?"

I fully endorse working toward an *eventual* goal of removing all nuclear weapons from the face of the earth. But that goal is like Robert Frost's distant star, "to stay our minds on and be staid," because history has shown that every weapon ever invented has been used in war. Pope Innocent II, in the year 39 A.D., declared the recently developed and deadly crossbow "hateful to God and unfit for Christians" and forbade its use. Only a few years later, this edict of the Second Lateran Council was amended to permit use of the crossbow against Moslems. Shortly thereafter, this limitation also broke down and the crossbow was used indiscriminately against one and all until more efficient means of killing superseded it. It is already a departure from the norm that for 44 years, since Hiroshima and Nagasaki, no one has employed nuclear weapons in actual conflict, although their use has been contemplated on numerous occasions.

With that distant goal in mind, we must address the physical facts and the practical problems of statesmanship. What can we do in the years that remain before the year 2000?

Let me start with two technical observations about nuclear weapons and stockpiles:

• Even if all nuclear weapons are destroyed or rendered harmless and all weapons fuel (fissile material) attenuated for use in power reactors, knowledge of how to make such weapons cannot be excised. Given a future commitment or urgency to restart making nuclear bombs, for whatever reason, the experience of World War II shows we are never more than a few years away from a nuclear weapons capability.

• There is no way to verify compliance with a total ban of nuclear weapons. The 1988 debate over ratification of the INF (intermediate-range nuclear forces) Treaty shows how hard it is to develop confidence that one can actually verify its

Sidney D. Drell, "Not So Fast," *Bulletin of the Atomic Scientists,* January/February 1989. Reprinted by permission of the BULLETIN OF THE ATOMIC SCIENTISTS, a magazine of science and world affairs. Copyright © 1989 by the Educational Foundation for Nuclear Science, 6042 S. Kimbark Ave., Chicago, IL 60637.

proscriptions against one class of weapons. We can see weapons being destroyed—and we could also weigh weapons fuel being contaminated if stockpile reductions were negotiated—but what about secret caches? Although we may be able to gain confidence that hundreds of weapons are not retained in a secret arsenal, I know of no way to get that number down to dozens in the near future.

Political Realities

Recognizing these two technical limitations, my argument is based on political realities: this is a world of nation-states, adversarial interests, fear and distrust between different economic and political systems, and intolerance grounded in religious beliefs. The removal of nuclear weapons must be tightly coupled to progress in reducing distrust, fear, intolerance, and in removing the image of the enemy between nations with adversarial relations. That is not to suggest that political and social progress must precede large reductions in our nuclear arsenals—or vice versa. Each can assist the other. Getting rid of nuclear weapons is much more difficult than a well-defined technical challenge. Albert Einstein once said, "Politics is much harder than physics."

"The political shape of a disarmed world is beyond my hopes for my children's and grandchildren's generations."

But the process can be started, and now is as good an opportunity as we have ever had. In two years the newly constructive dialogue between the United States and the Soviet Union achieved two notable results: the 1986 Stockholm accord on prenotification and observation of conventional military maneuvers and the Intermediate-Range Nuclear Forces Treaty.

One can hope that the pace will quicken as more constructive political relations and reduced fears lead to unilateral moves of good sense that need not await formal treaties and exquisitely balanced concessions. . . .

We are beginning to see the first benefit of new thinking calling for less threatening offensive force postures. Political accommodations and sensible unilateral moves to reduce and reconfigure conventional military forces may further accelerate progress.

We must greatly reduce the risk of nuclear war by seeking a stable strategic balance at a much lower level of nuclear forces. We should first reaffirm their purpose: simply to deter any use of nuclear weapons by an opponent. To this end, survivably based nuclear forces—at a fraction of their present level—are adequate.

But a major change—a phase change—in the world

political order will be required before we can abandon a policy of deterrence based on strategic offensive nuclear forces. The political shape of a disarmed world is beyond my hopes for my children's and grandchildren's generations. That is not a statement of despair. The young American republic of 1789, or again in 1812, could not very well foresee living in peace and fully sharing a *disarmed* 2,500-mile border with a nation of the British Commonwealth, yet that was achieved within a century. . . . I cannot at present identify all the practical steps leading to zero nuclear weapons—any more than I can to no weapons of any kind.

I can, however, see some useful forward steps:

● Negotiate more confidence-building measures to remove the threat of blitzkrieg conventional attacks.

● Alter deployed conventional forces to defensive postures where East meets West. Define realistic and essential security objectives and deploy and train forces to meet them without early reliance on nuclear weapons. In effect, make nuclear weapons less relevant.

● Strengthen international organizations; cooperate and devote resources to remove sources of conflict and to solve the more pressing of the earth's problems: hunger, ignorance, fear, disease, pollution, overpopulation.

● As scientists, assist the current efforts to achieve and enforce a treaty prohibiting chemical weapons. There is no way to verify total compliance with such a ban—just as there is no way to verify total nuclear disarmament. But, through active support and full cooperation in the self-policing of such a ban, we can help realize this goal. We can commit ourselves to it in a statement of conscience—a new Hippocratic Oath.

Efforts of Scientists

In response to revelations about the Libyan chemical weapons plant, West German Foreign Minister Hans-Dietrich Genscher urged the scientific community "to make their entire know-how available so that we can solve the still unsettled questions concerning a global ban on chemical weapons, especially the related verification issues."

Genscher's words were echoed by his Italian counterpart, Giulio Andreotti, who said that the problems posed by verification were difficult, but not impossible, to resolve: "Technical complexity cannot, and should not, be used as an alibi to delay the resolution of what is essentially a political problem, since it is based on the question of confidence between states," he said.

In this way scientists—who have created the technology of unparalleled means of mass, indiscriminate destruction—would begin to participate as a community in supporting declared national goals to ban such weapons. This step could move us toward what some day may be a nuclear-free world. . . .

I emphasized chemical weapons, first, because they are an immediate danger: they're being used. A million people were killed in the Iran-Iraq war, many through the indiscriminate use of chemical weapons. We've got to get rid of them quickly because the restraints have already broken down. A second reason has to do with the political aspects of the problem. Many countries have stated that they support banning chemical weapons. In that political framework we can do something, and we should make it an example from which to move on to the next, the nuclear problem. There's no other fundamental difference. It's not a matter of chemistry versus physics. But let's admit that one reason there is a political consensus to move on chemical weapons is because some who would otherwise be unwilling to ban them know that they have a nuclear fallback. So getting rid of all the "weapons of indiscriminate destruction" will be very difficult, but we have a possibility to eliminate one, and we should take that opportunity.

Unkept Promises

Other nations have every reason to give us hell at the next nonproliferation review convention. The nuclear powers have failed to live up to their promises. Just read the prologue, the first part of both the Partial Test Ban Treaty and the Non-Proliferation Treaty. We promised to work to reduce our nuclear forces, and to stop testing nuclear weapons underground—and we have done neither. There is no reason for other countries to look to us for leadership and to sit there and accept our fascination with continually making more and more accurate weapons while we say, "Don't do it yourself." We have to make clear that we share the goals of working to stop underground nuclear bomb testing at an appropriate time and working to reduce our arsenals to zero.

"Deterrence is not acceptable as an end in itself, but it is a morally acceptable policy, because it offers the best prospect of avoiding war now."

The difference is not with the goal but with the time scale. I am stuck in a phase similar to that of the American Catholic bishops in their landmark pastoral letter of 1983, reiterated in 1988: deterrence is not acceptable as an end in itself, but it is a morally acceptable policy, because it offers the best prospect of avoiding war now, so long as you're working toward gradual disarmament. I recognize all the inherent contradictions that people from Jonathan Schell to George Kennan have struggled with on the issue of deterrence. What sense does it make to threaten to use a weapon that you say no person of any humanity could conceivably use? Yet these

weapons exist.

There are two time scales in this world. There's the practical time scale of what we can do during the next decade, which is to reduce the number of weapons and to make deterrence more stable by working so that better command and control and survivable forces exist. There's a longer-term time scale that's beyond my technical vision, beyond my vision of the politics of the world, and calls for getting rid of nuclear weapons in the long run.

Sidney D. Drell is a physicist and arms control specialist. He is also the codirector of the Stanford University Center for International Security and Arms Control in Stanford, California.

viewpoint 4

U.S. Nuclear Policy Is Adapting to Soviet Reforms

James A. Baker III

The contrast in relations between Moscow and Washington in October 1962 and in October 1989 could not be greater. In October 1962, we faced a blustering Soviet Union; its leader talked of burying us. Today we face a sobered Soviet Union; its leader talks of restructuring his society.

In October 1962, the Soviet economy was growing and ready to feed an unrelenting arms buildup. Today the Soviet economy is virtually bankrupt.

In October 1962, the Soviet space program raised fears we would lose the race to the Moon. Today the Soviet Union is racing to avoid being left behind as much of the world moves from the industrial age into a new century.

And in 1962, we stood—as you all no doubt recall—eyeball-to-eyeball on the brink of war. Today, by contrast, superpower relations are as promising as we have ever found them since the Second World War. Looking back the Cuban missile crisis posed the clearest possibility for nuclear war in the postwar era. Looking forward we face the clearest opportunity to reduce the risk of war since the dawn of the nuclear age.

The President has described our purpose as moving beyond the peace of armed camps to the peace of shared optimism. I have previously described our strategy for achieving this goal through a prudent search for points of mutual advantage. Today I want to talk in more detail about one of those points: arms control.

Political Differences

Arms control can lend a strong hand in building an enduring peace, but arms control does not proceed in a political vacuum. Let me be clear: We compete militarily because we differ politically. Political disputes are fuel for the fire of arms competitions.

Only by resolving political differences can we dampen the arms competition associated with them. To follow Clausewitz, if war is the continuation of politics by violent, military means, arms control is the search for a stable, predictable strategic relationship by peaceful, political means.

That is why our times are now so full of promise. Over the last 40 years, arms control played only a limited role in shaping the U.S.-Soviet security relationship, because our political differences were simply too wide to allow enduring and substantial progress. Western strength and Western unity sustained deterrence throughout this period when we all lived in the shadow of opposed values and conflicting purpose. Now *perestroika* in Soviet domestic and foreign policy could, in part, lift the shadow. The political prerequisite for enduring and strategically significant arms control may finally be materializing. Surely the President was right when he wrote President Gorbachev, "We bear enormous—and mutual—responsibility to take advantage of the promise of these extraordinary times to improve international security."

The President and I have both said that we want *perestroika* to succeed. It would be folly, indeed, to miss this opportunity. Soviet "new thinking" in foreign and defense policy promises possibilities that would have been unthinkable a decade ago, such as deep, stabilizing cuts in strategic forces and parity in reduced conventional arms in Europe. Yet *perestroika*'s success is far from assured. Any uncertainty about the fate of reform in the Soviet Union, however, is all the more reason, not less, for us to seize the present opportunity. For the works of our labor—a diminished Soviet threat and effectively verifiable agreements—can endure even if *perestroika* does not. If the Soviets have already destroyed weapons, it will be difficult, costly, and time consuming for any future Kremlin leadership to reverse the process and to assert military superiority.

Reprinted from James A. Baker III, "Prerequisites and Principles for Arms Control," *Department of State Bulletin,* December 1989.

And with agreements in place, any attempt to break out of treaties will serve as one indicator of an outbreak of "old thinking."

We can take advantage of the new political climate to consolidate deterrence at lower levels of risk. Through sound and verifiable agreements, we can shape and institutionalize a more stable, predictable strategic relationship. The changing political relationship between the Soviet Union and the United States should be reflected in changing Soviet force structures and strategic concepts. In this way, we can help to codify political progress in military reality and by doing so, underpin that progress and strengthen it.

The Changing Strategic Environment

Before outlining the tenets of this Administration's arms control policy, I would like to say a few more words about the broader strategic environment in which arms control must operate.

Politically the Soviet Union is in the midst of this revolution of *perestroika, glasnost,* and democratization. The new thinkers understand that Stalin's system must change fundamentally if the Soviet Union is, as Mr. Gorbachev has said, to enter the 21st century in the manner worthy of a great power. To this end, the Soviet leadership has done much and promised even more for political, economic, and legal reforms. While his reforms need to be extended, codified, institutionalized, and made habitual, the political face of Soviet power is being changed already.

The prospects for reform are just as great—in some cases perhaps even greater—in Poland, Hungary, East Germany, Czechoslovakia, and the other countries of Eastern Europe. While the trends should not be overstated, the political foundations of a Europe divided by force since 1945 are crumbling away. We can move toward the President's vision of a Europe whole and free.

These great political changes are set in a time of vast technological changes. Our military tools are being reshaped by emerging technologies that could offer greater security. Advances in sensor technology, data processing capabilities, and precision-guided munitions present novel ways to strengthen deterrence.

We need to be careful, however, also to see the darker side of changing technological realities. More nations are acquiring the capacity to make chemical weapons and to manufacture missiles. With many of these regimes locked in continuing regional conflicts, the explosive escalation potential of their disputes is obvious.

Strengthening Deterrence

I would add, too, that these technological changes are taking place in a time of changing defense economics. Everyone has noted the Soviet Union's compelling need to convert some of its vast

expenditures for the military into domestic reconstruction. The era of rapidly rising defense budgets is over in the West too. From the new technologies, we are going to have to pick very carefully those weapons that strengthen deterrence most cost-effectively.

What do these political, economic, and technological changes add up to? Strategically, the world we've planned for since the Cuban missile crisis is increasingly distinct from the world we actually face. Threats to our interests are changing politically and multiplying technologically. Our capabilities are being improved technologically but constrained economically.

Our fundamental values and interests will endure. But as our strategic environment is transformed, we need to look anew at some of our guiding concepts and approaches. Many long-held assumptions may need to be rethought. Strategy aligns ends and means. As both shift, strategy may have to shift too.

"A coherent, integrated strategy . . . reduces the risk of war by deterring aggression while promoting American values."

For example, we need to think about the future of both European security relations and the central superpower strategic relationship. Today's historical political transformations in Eastern Europe—if suitably institutionalized—make such reassessments doubly important. In light of the growing threat to our global interests and power projection forces posed by the proliferation of new technologies, we also need to reconsider our strategy for Third World conflicts. Over the longer term, we need to consider if strategic defense options, deep reductions in nuclear and conventional weapons, increasingly powerful conventional munitions, and shifts in Soviet strategy will alter our requirements for deterrence.

Deterring Aggression

To cope with this changing environment, defense programs and arms control must work together. This is a prerequisite for a coherent, integrated strategy that reduces the risk of war by deterring aggression while promoting American values. Both defense programs and arms control can serve the common goals of enhancing stability, ensuring predictability, and bolstering deterrence. As our strategy may change in response to an evolving strategic environment so, too, our defense programs and arms control positions would also change. Together security will be enhanced.

Clearly neither defense programs nor arms control can do the job alone. No remotely achievable START [strategic arms reduction talks] agreement, for example, can restore the survivability of our silo-based ICBMs [intercontinental ballistic missiles]. To maintain the integrity of the triad, we will need to rely upon the deployment of mobile missiles as a key component of our nuclear modernization program. But START can play a key role. It can reduce the Soviet threat to our forces and thereby make survivability through mobility more feasible. Without START to constrain the Soviet threat, the job of ensuring reliable deterrence would be less predictable and affordable. Without the START negotiations, the domestic consensus needed to support essential modernization programs—not only mobile ICBMs but also B-2, Trident, and SDI [Strategic Defense Initiative] would be difficult to sustain. Likewise without our strategic modernization program, the benefits of a START agreement would be sharply reduced. Thus our force modernization and arms control efforts reinforce one another.

In September 1989 I announced the President's decision to allow mobile land-based missiles in START. Permitting mobile missiles only makes sense if the United States is willing to deploy them. For this reason, this decision is contingent on congressional funding of our mobile missile program. Congress needs now to support START, not undercut it, by funding this program. As Senator [Sam] Nunn said, "Unless we in the Congress can manage to put our ICBM modernization program back on track . . . the START negotiations face a very bleak and a very long future indeed."

Another prerequisite for a successful strategy—for defense programs and arms control that work together—is the need for unity as a nation and as an alliance. This follows from a simple truism: United we stand, divided we fall. We should not tempt the Soviets with exploitable differences between the Administration and Congress or between the United States and its allies. That does not exempt us, of course, from the need for informed debate. It is imperative that we maintain open and honest discussions about strategy and arms control matters within the strategic community and with the public at large. As we deter possible aggression, we must—as the noted military historian Michael Howard has put it—reassure our peoples that their defense dollars are efficiently and effectively supporting the cause of peace. An open, frank debate is the surest formula for unity. But such a debate must begin and seek to reach some resolution before treaties are signed if we are to bring home treaties in the national interest.

The Goal of Arms Control

As a contribution to such a debate, I would like to move now from the prerequisites of arms control to the basic goal of our arms control policy and the principles for achieving it.

The main goal of arms control is to reduce the risk of war—any war, nuclear or conventional. We hope to prevent war by working toward a stable, predictable strategic relationship. Stability requires military forces and policies such that no one can gain by striking first even in the worst crisis. Beginning a war, especially a nuclear war, must never become a Soviet option—even a least-worst option, as a noted strategist once put it. Predictability requires that sufficient openness and transparency prevail to prevent misperception, miscalculation, and an inadvertent war—a war no one wanted but no one could stop. The more open and transparent Soviet military affairs, the greater trust and confidence we can have in Soviet intentions.

"Greater openness is the surest path to greater predictability and a lower risk of war."

Four principles guide our search for a stable, predictable strategic relationship.

First, we seek reductions in first-strike, surprise attack capabilities. We seek stability through proposals to reduce those capabilities most suited for offensive, *blitzkrieg*-style actions and preemptive first-strikes. In CFE [conventional armed forces in Europe negotiations], we've concentrated on eliminating Soviet advantages in those weapons most suited to seizing and holding territory: tanks, artillery, and armored personnel carriers. In START we have focused on reducing the most destabilizing weapons, especially vulnerable, silo-based heavy ICBMs, such as Soviet SS-18s. These weapons are suited principally for preemptive first-strikes and not for retaliatory missions. In Wyoming we proposed banning short-time-of-flight sea-launched ballistic missile (SLBM) tests, seeking in this way to reduce the capability for a Soviet decapitating first-strike. Our START proposals emphasize the relative merits of slow-flying weapons—such as cruise missiles and bombers which are not suitable for a first-strike.

Our SDI program also supports our emphasis on stability. Effective strategic defenses can contribute to survivable, cost-effective barriers to a successful first-strike. That is why we look favorably on the decision made by the Soviets in Wyoming to delink the defense and space talks from START. This Soviet decision to no longer hold START hostage to resolution of defense and space issues removes a key obstacle to a START treaty while enabling us to proceed with our SDI plans. We remain committed to preserving our right to conduct SDI activities consistent with the ABM [Antiballistic Missile] Treaty. And we will use the defense and space talks to explore a cooperative and stable transition to a greater reliance on stability-enhancing, cost-effective strategic defenses.

Our *second* principle—predictability through openness—expands the traditional focus of arms control on capabilities. Every war has its own unique causes, but surely Thucydides made an important general point when he wrote, "What made war inevitable was the growth of Athenian power and the fear which this caused in Sparta." Arms control has mainly focused on the first part of this equation: constraining or reducing destabilizing military capabilities. Now in expanding the agenda, we are working to deal with the other aspect of Thucydides' equation: fears of aggressive intent. We are pushing to make Soviet military activities more open and transparent. The more we know and understand, the more we can be assured that our fears are not results of misperception or miscalculation. Greater openness is the surest path to greater predictability and a lower risk of war, especially inadvertent war.

The President's "open skies" initiative is a clear example of this new focus in arms control. Openness about military forces and activities is at the heart of the talks on confidence- and security-building measures (CSBMs) among all the states of Europe. In those negotiations, we are proposing an all-European military data exchange about our forces and weapons programs in keeping with the spirit of openness we found at the Wyoming ministerial, we signed an agreement on notification of strategic exercises and invited the Soviets to visit our SDI facilities. The chemical weapons data exchange will help us move toward a verifiable global ban. [Soviet] Defense Minister Dmitri Yazov's visit in October 1989 is just one of a series of exchanges that provide face-to-face opportunities to understand the Soviet military. And we have pushed the Soviets to publish a real defense budget that reveals the inputs into and outputs from their defense production process.

"Effective verification can ensure that the treaties we sign are doing their job to institutionalize a safer world."

Openness in military affairs is just part of our overall emphasis in our dealings with the Soviets on creating open, pluralistic institutions. On his visit, Soviet Defense Minister Yazov talked of the increasing influence of Supreme Soviet committees over the Soviet defense complex. We hope that Soviet military power may increasingly be exposed to the salutary effects of detailed and searching public debate.

Greater openness combined with force reductions will support political change as well. In CFE our proposals will reduce the potential not only for a Soviet *blitzkrieg* but for Soviet intimidation of Western Europe. The Soviet Army we face as a potential army of aggression is to East Europeans an army of occupation. The weight of the Soviet military presence in Eastern Europe will be reduced. Freed from the cold shadow of Soviet military domination, political pluralism and free markets should flourish more easily in Eastern Europe.

A more predictable strategic relationship should also be less expensive. Arms control can, as the President wrote Mr. Gorbachev, "introduce predictability into military planning so that we can slow the pace of military competition." A slower competition could be a cheaper and safer competition. But our desire to save money must not come into conflict with the necessity for security.

Nuclear Proliferation

The *third* principle of our policy is a broadened arms control agenda, far wider than its traditional East-West nuclear focus. We are broadening our agenda with the Soviets, both in terms of dealing with pressing global arms control problems, like chemical and missile proliferation, as well as focusing on regional conflicts. In an increasingly intertwined world, a stable, predictable U.S.-Soviet strategic relationship depends in part on regional stability and vice versa. Earlier I noted that advanced technologies were proliferating to the Third World. Advanced fighters have gone to Libya, Syria, and North Korea. Over 20 states possess the capability to produce chemical weapons. And nuclear proliferation, notably North Korea's reactor program, remains dangerous. Arms control should increasingly focus on such problems.

The President's UN [United Nations] initiative can lead us toward a verifiable global ban on chemical weapons. The President's proposal represents a realistic road map for progress. As a step toward a multilateral ban, we will move bilaterally with the Soviets to reduce chemical weapons to 20% of the current U.S. levels. We will further slash stocks to just 2% of their current levels within 8 years after the multilateral convention goes into effect. This total cut of 98% is a substantial acceleration of previous destruction plans. Then we will move to zero within 2 years of adherence to the ban by all chemical weapons-capable states.

We realize it may be difficult to persuade problem states such as Libya and Iraq to join, but we are creating an environment where everyone will have incentives to join and costs to pay for remaining an outlaw. Export controls on precursor chemicals will be strengthened, building on progress made at the Canberra conference. The President has also ordered a study on sanctions to deter and punish chemical weapons use and other violations of a convention. States must know that they will pay a price for their inhumanity.

Our *fourth* principle is institutionalization of a safer world. The President aims to reduce the risk of war permanently, not temporarily. We want to see Soviet defensive military operations made habitual. We want to see the "new thinking" concretely built into the Soviet force structure. We want to see weapons destroyed, not merely removed. And we want agreements that can endure.

Effective verification can ensure that the treaties we sign are doing their job to institutionalize a safer world. Because of the primacy of effective verification in this Administration's approach to arms control, our negotiators have already proposed data exchanges and trial verification measures that would be implemented even before the agreements themselves are concluded. Such measures in START and in chemical weapons will help us build confidence and gain practical experience that will facilitate the conclusion of sound, verifiable agreements. A sustainable and enduring arms control process also means avoiding limits, for example, on sea-launched cruise missiles, that would create unmanageable verification and compliance problems.

Neither have we stood still in pressing the Soviets to comply fully with agreements already signed. In September 1989 President Gorbachev informed the President that the Krasnoyarsk radar would finally be destroyed. We welcome Moscow's step to come into compliance with the ABM Treaty.

A Realistic Path to Reduction

These four principles of a more stable, open, broader, and less reversible strategic relationship offer a realistic path to a lasting reduction of risk. It is a path best traveled by steady steps that build on one another, rather than grand leaps that are often as not unrealistic or undone. In START, in CFE—in all our negotiations—we have made fair, responsible proposals designed to find enduring points of mutual advantage. The Soviets have said yes to much of what we have proposed. Now we have rolled up our shirt sleeves and set to work together to put principle into practice.

"We cannot disinvent nuclear weapons nor the need for continued deterrence."

We should be clear about the task ahead. We are not on the verge of a perpetual peace in which war is no longer possible. We cannot disinvent nuclear weapons nor the need for continued deterrence. Nor can we completely eliminate Soviet-American rivalry. But that rivalry does not require that we stand on the brink of Armageddon as we did in 1962. Peace need no longer hang solely on Winston Churchill's "process of sublime irony . . . where safety will be the sturdy child of terror and survival the twin brother of annihilation."

Deterrence need not rest only on a delicate, technical balance of terror disturbed by periodic crises. Opportunity invites us, instead, to move beyond containment, beyond the cold war, to a new strategic relationship based on a sound political footing.

A new relationship in which the capabilities and incentives to attack first are minimized and the possibilities of strategic defenses are pursued. A new relationship in which Soviet military power is open to the naked eye, not just satellites in the sky. A new relationship in which all the peoples of Europe are free of military intimidation. A new relationship in which effectively verifiable treaties lock in a lower risk of war. And a new relationship in which arms control aids our people in turning the seeds of war into the fruits of peace.

This is the strategic relationship we seek.

James A. Baker III is the Secretary of State for the Bush administration.

*"The . . . [U.S.] approach is too passive
toward the recent, and still continuing,
transformation of the Soviet empire."*

U.S. Nuclear Policy Is Not Adapting to Soviet Reforms

Fred Charles Iklé

From Berlin to Baku, popular revolutions are dissolving the world's last empire—the erstwhile evil one. Upheavals continue. The *annus mirabilis* of 1989 liberated political energy that is now beginning to affect every corner of the globe: hastening the demise of Beijing's outdated dynasty, eroding Fidel Castro's and Kim Il Sung's dictatorships, creating a new German nation, transforming the European Community, and perhaps draining the marrow out of NATO [North Atlantic Treaty Organization].

Routinely, Pentagon planners stake out their work each year with a description of The Threat. Now we see in astonishment that in every arena of confrontation The Threat is being turned upside down. Indeed, our arch-adversary's arch-alliance, the Warsaw Pact, remains strung together only by the thin filament of a vacuous treaty text, having lost its ideological glue and Stalinist discipline.

What, now, are the threats against which the Pentagon should prepare? How should America's strategy and military forces, indeed its overall foreign policy, be changed to take account of the transformed environment?

New Strategy Needed

In 1990, Congress will vote on the first defense budget since democratic revolution swept through Eastern Europe. The Bush administration sees this budget as beginning "the transition," in the President's words, "to a restructured military—a new strategy that is more flexible, more geared to contingencies outside of Europe. . . ." But as the defense budget winds its way through Congress, one must fear it will be treated like a big sugar loaf from which to shave off sweet slices: cut more army divisions here, lop off another aircraft carrier there, cancel the new strategic bomber, cut strategic defense by half, and so on—chop, chip, chop.

One can no more construct a new strategy from canceled defense programs than one can build a house from woodshavings. Alas, any sense of urgency in Washington that has now welled up about defense is aimed at budget cuts, not grand strategy. Stubborn fiscal pressures provide today's motive for changing our military programs and forces. As for America's overall strategy, influential voices in the administration and in Congress maintain that because of vast uncertainty in the world we should change warily. The United States would be rash, it is argued, if it sought to shape the ongoing transformation of the global strategic structure, a transformation that is not only unpredictable but, in any event, largely beyond our influence. A renovation of our security strategy, according to the conventional wisdom in Washington, will have to wait till the dust settles.

This complacency is mistaken. For we should care immensely just how "the dust settles."

To say that the United States can wait to address the fundamentals of our Western security strategy is both too complacent and too unambitious. It is too complacent about the potential losses, the utter disaster that the fragility of today's global transformation could bring. And it is too unambitious and too passive, given the potential gains, the promise for enduring and profound improvement in our security that these pregnant times hold. Both the complacency and the listless passivity stem from a poverty of imagination about the potential for change— indeed, about the impact of the changes that have already occurred. What constricts our imagination are old habits of the mind, an almost unwitting reliance on the strategic concepts that have shaped the ends and means of our defense policy for decades.

The most influential of these concepts have to do with Europe, precisely the area that has now experienced the greatest change. For more than four

Fred Charles Iklé, "The Ghost in the Pentagon," *The National Interest*, no. 19, Spring 1990. © 1990 *The National Interest*, Washington, D.C. Used with permission.

decades, year after year, the threat of a massive Soviet invasion of Western Europe has determined the design and purpose of over half of America's resources for defense.

"Washington's national security establishment continues to see the world in terms of the 1947 mindset."

As pervasive as it is obsolete, this mindset took hold forty-three years ago. In 1947, just two years after the Second World War, planners of the Joint Chiefs of Staff set down some views of a possible war with the Soviet Union. "The Soviet land armies and air forces are capable of overrunning most, if not all, of Western Europe in a short time," warned their memorandum. "The ability of the Allies to meet and retard the Soviet efforts would depend to a very large degree upon the length of the period of warning they receive and the use they made of it." Gloomily, the assessment listed what must be done "if the war is not to be lost." The principal requirement, it argued, is to prepare for "an offensive strategic air effort against vital Russian industrial complexes and against Russian population centers," [argued by Thomas Etzold and John Gaddis].

All these concepts have survived to this day. The whole mindset is there: the Soviet military threat to the center of Europe, U.S. dependence on warning time, and—to compensate for Western weakness —U.S. strategic bombing of Russia. Like a sturdy genetic code, the mindset propagated itself through generations of technological revolutions in armaments; through the Korean and Vietnam Wars; through the Sino-Soviet split and the build-up of British, French, and Chinese nuclear arsenals; through the economic empowerment of Japan and the growing unity and economic expansion of Western Europe.

Let us defer to another day the question whether this concept of the dominant threat remained valid beyond the mid-1950s; that is to say, beyond the death of Stalin, the rebuilding of Western Europe's economies and military forces, the consolidation of NATO, and the massive expansion of U.S. nuclear might. But surely, during 1989 a few more things have changed in the center of Europe—and indeed, in Moscow.

Nonetheless, Washington's national security establishment continues to see the world in terms of the 1947 mindset. By regarding the basic strategy as an unchanging core, it recognizes improvement only at the edges. It admits—grudgingly—that the Warsaw Pact's massive attack on Western Europe would be somewhat weaker now and preceded by additional warning time. Having figured out that the attack would be weaker, the Pentagon bureaucracy

concluded it could still adhere to the same old strategy, even if it sacrificed two army divisions and a few tactical fighter wings on the budget-cutting altar. Second, having discovered increased warning time, the Pentagon bureaucracy now accepts converting some of our forces intended for Europe from active status to the less costly reserve status. While the shift to reserves has merit, how the bureaucracy rationalizes it is preposterous. It totally misses the import of the revolution that has swept through Eastern Europe to conclude merely that we can trim our active forces a bit because our intelligence experts now promise, say, thirty-seven days of warning (instead of fourteen) in the event of the Second Coming of World War II. Since 1989, NATO's warning time is to be measured neither in fourteen days nor in thirty-seven days, but in years—the years it would take to re-Stalinize Eastern Europe.

Even while Moscow still controlled Eastern Europe, the concept of a fixed warning time measured by a given number of days was mistaken. Advance warning of an attack is almost always ambiguous, and a clever enemy will use every means of deception to deepen ambiguity. It is hard to believe that the United States and its European allies could ever agree, in response to such warning and before the first shot was fired, to the planned deployment of six additional U.S. divisions to Europe.

By postulating that the hoary scenario of the Warsaw Pact onslaught comes packaged with an extra twenty-three days of "warning," the Washington establishment justifies cuts in the defense budget. By trimming the edges of our force posture—cut some divisions here, eliminate some bases there—it can accommodate some further budget cuts. Most of these cuts will be welcomed by most members of Congress—unless and until they hurt vociferous constituents. But the totality of this approach is too passive toward the recent, and still continuing, transformation of the Soviet empire. It fails to cultivate and to secure the hitherto unimagined gains for peace and democracy.

Undermining Gains

Not to worry, say some U.S. officials, these gains will be secured through arms control agreements, particularly through the talks in Vienna on conventional force reductions in Europe. While a successful conclusion of these talks will undoubtedly bring security benefits, their most lasting impact might yet be to undermine some of the recent gains for Eastern European independence. The talks in Vienna are aimed at a treaty that will enshrine the concept of parity between NATO and the Warsaw Pact and impose the same limits on American forces in Western Europe as on Soviet forces in Eastern Europe. To be sure, President George Bush's recent proposal to lower these limits and to move away from equality between American and Soviet forces would improve the outcome. Yet, most of the new governments in Eastern Europe are now seeking agreements with Moscow for the withdrawal of *all* Soviet forces from their countries. When these governments later sign the grand arms treaty in Vienna, they will in effect legitimize a Soviet occupation force in their countries that their new bilateral agreements with Moscow will have happily eliminated.

Why should the United States and its NATO allies labor in Vienna to give birth to a treaty that will help maintain the Warsaw Pact as a coequal to NATO? Why should the Pentagon bureaucracy still allocate over half its resources to a conventional war in the center of Europe against a Warsaw Pact invasion with only thirty-four days (or whatever) of warning? The answer to both these questions is the same: our arms control policy and our arms policy are dominated by the same obsolete mindset.

The forty-year-old image of The Threat and our forty-year-old strategy constrict our capacity to grasp the immensity of the global change now unfolding before us. We have promoted a "stable balance" between NATO and Warsaw Pact forces for so long that quite a few Westerners have come to think Soviet forces in the center of Europe are needed for the sake of stability. One even hears whisperings in some mossy NATO circles that the Warsaw Pact ought to be preserved to maintain this treasured stability.

The Two Germanies

In London and Paris, moreover, some officials still pine for a permanent partitioning of Germany; not only to preserve the "stable" East-West balance, but also because they begrudge Germany its growing strength. They try to disguise this envy as legitimate, psychic trauma from both world wars. One wonders, though, if Paris must fear a new invasion by Hitler or the Kaiser, should Bonn fear a new Napoleonic war? Were it not for the democratic vigor of the people in the Eastern and Western part of Germany, these British and French stability worshippers might yet succeed in rebuilding the Berlin Wall. As Dr. Samuel Johnson might say today, "Stability is the last refuge of a reactionary."

Again, those who want to "stabilize" the military confrontation between NATO and the Warsaw Pact are, in one sense, too unambitious. Beholden to the old strategic mindset, they eschew improvements in our security that an up-to-date strategy could achieve. In another sense, more dangerously, they are too complacent in believing that NATO could protect itself against a "second Stalin" simply by manning its old fortresses again.

Prudently, our defense planners are worried by intelligence information showing that the Soviet Union, despite its acute economic and nationality crises, continues the mass production of a wide array of powerful armaments, including the most modern nuclear missiles. Our planners would be remiss if

they failed to project these formidable military capabilities into a less benign political context than we enjoy today. We might wake up some morning to find a new aggressive dictatorship in Moscow, a ruthless tyrant who could order Russian forces to reconquer Eastern Europe and menace NATO more dangerously than Stalin ever did.

"If Paris must fear a new invasion by Hitler or the Kaiser, should Bonn fear a new Napoleonic war?"

But what a fatal error to believe that, in the event of such a catastrophe, NATO could simply pick up where it left off in 1988! It is grossly unrealistic to assume that our Atlantic Alliance would proudly reassemble at the old ramparts, regenerate its military exercises in West Germany, deploy new short-range nuclear arms in Europe (as had been planned in 1988), and induce its member nations to increase their defense budgets again.

If the will of all the peoples in Eastern Europe had been crushed under Soviet tanks, if the democratic forces from Sofia to Warsaw had been drowned in rivers of blood, if the profound German aspirations for a unified nation were cruelly affronted by new mine fields and walls, if the expectations for a peaceful, open continent now animating all the nations of Europe were totally shattered—how could NATO then return to "business as usual"? Instead, the Atlantic Alliance would be rent by harsh recriminations. Its governments and its people would have lost confidence in the old strategy; they would recoil from the prospect of another forty years of military confrontation in the center of Europe. With so dark a future, the now ebullient spirit of the European community would falter. The fear of nuclear war would again weigh heavily on the public psyche, a fear the enemy could easily exploit to stir up disunity within the alliance.

The conventional arms agreement would offer scant protection in such a calamity. To be sure, thanks to the agreed reductions Russia would enter the new confrontation without its former advantage in numbers of troops, tanks, and other equipment. But by violating the agreement while NATO debates what to do, Russia could get a head start. Moreover, once Russia began its reconquest of Eastern Europe, the Vienna conventional arms agreement would be washed away like a sandcastle by the tide.

Changing Priorities

Again it becomes apparent here that our old strategic mindset imposes mistaken priorities on the renovation of America's (and NATO's) defense effort and on the West's arms control policy. Among the top priorities of our defense policy today ought to be the protection and consolidation of the recent political gains in Eastern Europe, and the removal of temptations among Russian "hard-liners" and would-be Stalins to use military force for a new expansion of the empire—all the way into the center of Germany, to the Adriatic, into Afghanistan, and beyond.

"We must make clear to the leaders in Moscow . . . that the West would never accept a new subjugation of Eastern European nations."

This means we must help to hasten the dismantling of the Warsaw Pact and the withdrawal of Soviet forces from countries where they are not welcome, even at the price of more drastically and more rapidly reducing U. S. troops deployed in the Federal Republic of Germany. On this point, the President's newest proposal moves in the right direction. Of course, we can hope and reasonably expect that a certain presence of American forces east of the Rhine will, for the foreseeable future, be welcomed by the German government and by a majority of the German people. This does not mean we should insist that a united Germany must, in its entirety, be a full-fledged member of NATO. But a residual, continuing U.S. troop presence in Germany would symbolize and give reality to a security link between the United States and Europe, thus complementing the enduring spiritual and cultural bonds of the Atlantic Alliance. To preserve a transatlantic security link is an imperative for the long-term benefit of the world.

We can best preserve the global benefits of the Atlantic Alliance by taking the initiative now, in concert with our allies, to shape a new security system for Europe. Instead of husbanding our military assets to "stabilize" the NATO-Warsaw Pact confrontation in the center of Europe, we should shift more resources and effort to help stabilize democracy in Eastern Europe. It is a mistake to wait for the Vienna arms reduction talks to establish such a system. These talks will wind up preserving the Warsaw Pact like a toad in a bottle of formaldehyde.

Fortunately, with every passing month, democracy in Eastern Europe is becoming stronger. Nonetheless, it is conceivable that some new crisis might suddenly tempt Moscow to consider military intervention in Poland, East Germany, or Czechoslovakia, much as Brezhnev stumbled into the decision to invade Afghanistan. *To deter such a decision, under any and all circumstances, is a mission of our national security policy that deserves much higher priority today than the forty-year-old Pentagon mission of deterring the now exceedingly improbable Russian invasion of Western Europe.*

To this end, we must make clear to the leaders in Moscow, whoever they may be, that the West would never accept a new subjugation of Eastern European nations. This means breaking with some shameful past precedents of American indifference. For example, only a few weeks after Soviet tanks crushed the democratic uprising in Hungary in 1956, President Eisenhower stealthily signaled to Moscow that the United States wanted to return to business as usual, despite our vehement public denunciations of the Soviet actions in Budapest. Similarly, after Brezhnev's invasion of Czechoslovakia in 1968, President Johnson's first priority in East-West relations was to resume the strategic arms talks. Prevented from doing so by the incoming Nixon administration, he felt great disappointment; by contrast, he had been much less troubled by Brezhnev's destruction of Czechoslovakia's democratic forces. We now know that this intervention set back democratization in Eastern Europe by twenty years.

Brezhnev's invasion of Afghanistan in 1979 provoked a more coherent and, above all, a more persistent American response. The Carter administration started and the Reagan administration greatly expanded military support for the Afghan resistance. Nine years later, the tenacity of this resistance—and, thanks largely to our help, its effectiveness—compelled the withdrawal of the Soviet forces. Moscow learned a lesson that must have weighed heavily on its decision to abandon its imperial expansionism. If Moscow is not to forget this lesson, the West must continue to remember it as well.

The Pentagon and Afghanistan

A lesson within this lesson is particularly relevant for the Pentagon's current adjustment to the transformation of the Soviet empire. The Pentagon bureaucracy opposed the Reagan administration's efforts to provide more effective weapons to the Afghan resistance. It held no grudges against the Afghan freedom fighters, of course; it merely wanted to save its weaponry (even some of the oldest models) to meet The Threat it knew since 1947, with its anticipated huge tank and air battles in Germany.

Specifically, the U.S. army bureaucracy fiercely opposed giving the Afghan resistance Stinger missiles, the hand-held surface-to-air missiles with which the Afghans could shoot down Soviet aircraft. Only a minimal fraction of the U.S. army stocks of the oldest version of this missile was needed, and once made available, turned the fortunes of the war in Afghanistan decisively against the Soviet invader. While opposing this small contribution, the Pentagon was unstinting in shoring up the ramparts against a Soviet invasion of the Persian Gulf. That there might be a connection between Soviet success or failure in subjugating Afghanistan and Moscow's appetite for invading Iran, Kuwait, and Saudi Arabia was not apparent to those who jealously guarded the hoard of thousands of old Stinger missiles.

Would it be unfair to see in this story a parable for today's need to set the right priorities between saving assets for World War III and helping to strengthen the forces for freedom now? Many members of Congress and private foreign policy experts are advocating larger and swifter funding for a whole panoply of reconstruction assistance to Eastern Europe—support for new private agriculture, management training for business and government. Clearly, if democracy can be firmly anchored in these countries the security of our European allies will be immensely improved. Efforts by the U.S. government to this end, hence, could reasonably be regarded as complementing or substituting for other U.S. efforts on behalf of NATO.

"Much as the 'Second World'—the Soviet empire—has changed . . . we must expect to see surprising changes and major geostrategic transformations in the rest of the world."

The Pentagon, however, tends to resist such a trade-off; and so might, when push comes to shove, various congressional committees. Despite the prudently chosen cuts that Defense Secretary Richard Cheney proposed to Congress for the 1991 budget, about half our defense effort is essentially still devoted to fighting a massive conventional war in Europe. Before cutting our NATO-related forces further, the Bush administration wants a signed agreement to bring Soviet strength in Eastern Europe down to ours in Western Europe—clearly a vast improvement in terms of military force ratios. But if democracy should not survive in Eastern Europe, no piece of paper signed in Vienna would offer us worthwhile protection.

Third World Contingencies

Over time, of course, the Pentagon will try to redesign its conventional forces for contingencies other than a war against a massive Warsaw Pact attack. The newly fashionable arena for our army, navy, and air force planners is the Third World. All types of armaments have suddenly become "vital" for dealing with Third World conflicts. Undoubtedly, each type has considerable military merit—whether it is the new C-17 transport plane that the air force wants, the new V-22 tilt-rotor plane desired by the marines, or the aircraft carriers treasured by the navy.

Four points need to be kept in mind concerning this now fashionable focus on military contingencies in the Third World. First, for the Pentagon, the problem of Third World hostilities is not new at all. Every war in which the United States has been involved since World War II—either directly with its forces or indirectly by providing military aid—occurred in the

so-called Third World. (Korea in 1950 was still a Third World country.) Since the war in Vietnam, American defense experts have conducted innumerable studies of the weaponry and tactics for use in Third World conflicts.

A second point of importance is that the concept of "Third World military conflicts" covers many different contingencies with vastly different circumstances in terms of strategic geography, types of forces, and weaponry used. To be prepared for such disparate contingencies, the United States needs to be able to rely on a wide array of different armaments. The Pentagon's planning, hence, needs to address many different possible threats in Third World situations. Some types of equipment that our military services are wont to neglect may well be essential for some of them; for example, devices to clear land mines in insurgency warfare, ships equipped to clear sea mines, drones for intelligence collection, and other types of unmanned air vehicles.

A third significant factor is the dominant role of military assistance. Should the United States have a major stake in a future Third World conflict, chances are it would seek to rely on assisting its friends, rather than engage its own combat forces. This prospect renders all the more urgent the long overdue overhaul of the awkward patchwork of laws governing military assistance, a task that Congress is reluctant to take up.

A Challenging Future

A fourth point needs to be stressed, although it may not be welcome in the Pentagon. Only a fraction of the Pentagon budget—less than a third—can be justified in the foreseeable future by the need to prepare for possible U.S. involvements in Third World hostilities. Even though some of the Third World countries are heavily armed, their arsenals are still small compared to the Soviet threat to Western Europe against which we have been preparing all these years.

Yet, much as the "Second World"—the Soviet empire—has changed to a degree and with a rapidity that almost no one had foreseen, we must expect to see surprising changes and major geostrategic transformations in the rest of the world. This potential for revolutionary change confronts Pentagon planners with a tough challenge. The ratio of the speed of political and diplomatic transformations—revolutions, alliance shifts, imperial expansion, and disintegration —to the speed of weapons development and procurement is about ten to one or even thirty to one. To complete the research and development of a modern weapons system takes ten years or more, to build and deploy it another ten years, and once deployed the system may remain in our forces for another thirty years. Who can foresee our strategic requirements for half a century?

This epochal time span has particularly frightening implications for nuclear weaponry. Our nuclear strategy is still under the curse of Joseph Stalin. Few realize the extent to which the design and purpose of our nuclear armaments, doctrine, and war plans date from the same old mindset that since 1947 shaped and governed the bulk of our conventional forces.

In that Stalinist era we sought to deter the Red Army from marching to the English Channel by threatening to drop atomic bombs on Moscow and on Stalin's war industries. Once we had acquired a great many more nuclear weapons, and once the Soviet Union deployed nuclear bombers and missiles, the top priority for the U.S. Strategic Air Command became the destruction—the instant the Red Army crossed the West German border—of as many Soviet bombers and missiles on the ground as possible. Since then, the concept of such a prompt, all-out strike has become a dogma that warps the design of our strategic forces to this day, even though it had become impossible to disarm the Soviet nuclear forces with such a strike many years ago.

The obsolete dogma that our nuclear retaliation must be prompt is responsible for the Pentagon's insistence that we must maintain a large force of land-based missiles, with all the difficulty and expense this entails. More dangerously, it perpetuates a vulnerable and hence a hair-triggered deterrent of thousands of missiles, both American and Soviet, sitting there like a thousand Chernobyls—till something, someday, goes dreadfully wrong. The confrontation of these U.S. and Soviet missile forces has evoked a morbid fascination among many defense technicians. By a banal and unrealistically abstract calculation—the so-called "missile exchange"—these technicians pretend to measure the "stability" of deterrence.

The Stalinist threat to Western Europe created other evil legacies for our nuclear strategy and forces. To prolong the life, or reach, of our nuclear deterrent against the feared Warsaw Pact invasion, we deployed a great many shorter-range nuclear arms in Europe, especially in Germany. All these nuclear artillery pieces, missiles, and nuclear-armed aircraft eventually, like a Sorcerer's Apprentice, acquired a life of their own. They became "vital," had to be modernized, and gave birth to a totally incoherent doctrine—Flexible Response—which flatly contradicts the "stability" doctrine of the "missile exchange."

Given the contradictions and shortcomings of these strategic concepts, perhaps the time has come to pay some attention to Soviet criticism of our nuclear deterrence doctrine. Gorbachev called mutual deterrence a source of tension. As Soviet foreign minister Eduard Shevardnadze put it, "nuclear deterrence inevitably perpetuates the totality of confrontational relations among states."

Flexibility Needed

Defense Secretary Cheney has wisely requested increased funding from Congress for certain research and development projects—in particular, strategic defense—that will purchase us flexibility in terms of doctrine and enemies. What we now develop and build will have to serve our military strategy in the twenty-first century. For the foreseeable future, one must hope, America's nuclear strategy will continue to be an alliance strategy embracing and protecting a non-nuclear unified Germany as well as a non-nuclear Japan—but a strategy that can put behind us the "confrontational" bipolar relationship with the Soviet Union to which Shevardnadze referred.

"Stalin has been buried twice in Moscow, but his ghost lives on in the Pentagon."

Yet before we wax lyrical about the dawn of a new era, free of the danger of instant nuclear holocaust, we have to remember that Stalin's legacy is not so easily overcome. The laws of physics, to be sure, do not ordain that there must be two nuclear superpowers, dividing the world into "two sides" threatening each other indefinitely with mutual annihilation. It is habits of mind and bureaucratic inertia, in both Washington and in Moscow, that cling to the apocalyptic "two sides" confrontation Stalin inflicted on the world at the end of World War II.

Such inertia casts a dark shadow far into the next century. The Pentagon bureaucracy continues to disparage strategic defense, contrary to the policy of the President and Secretary Cheney; it keeps designing our nuclear forces to deter a Warsaw Pact onslaught—and thus favors nuclear weapons installed in the middle of Germany and hair-triggered missile forces. Stalin has been buried twice in Moscow, but his ghost lives on in the Pentagon.

Fred Charles Iklé is a Distinguished Scholar at the Center for Strategic and International Studies in Washington, D.C. He was the undersecretary of defense for policy in the Reagan administration.

The U.S. Should Cut Defense Spending

Gene R. La Rocque

We Americans talk a lot about "national security." The term ranks right up there with the "Stars and Stripes" in emotional appeal. Outside the Pentagon few agree on what the term means, but everyone agrees we need it. Only the Joint Chiefs of Staff have actually defined the term national security in their dictionary of military terms.

Their definition emphasizes military superiority: "a military or defense advantage over any foreign nation or group of nations."

Over the past forty years this drive for security has taken on overwhelming military emphasis and now involves over nine million people actively engaged in the defense establishment at a cost of six billion dollars each week. It is this emphasis on the military which has sparked a growing number of appeals for a redefinition of the term national security.

The Center for Defense Information for many years has stated that "strong social, economic, political, and military components contribute equally to the nation's security." Thus we emphatically endorse the proposition that a strong, productive U.S. economy is central to the national security.

America's deteriorating international trade position is more damaging to our security than any new Soviet weapons development. The enormous burden of the rising federal budget deficit threatens the lives and prosperity of our children. The American educational system needs a major new infusion of creative ideas and resources if we are to retain our world influence in the future.

Future Military Needs

The most important factor which will determine our military requirements in the 1990's is who, where, and when the nation will call on the military to fight. The Soviet Union is the only nation, outside

Reprinted from Gene R. La Rocque, "National Security in the 1990s," *The Defense Monitor*, vol. 18, no. 8, 1989.

of Britain and France, which could destroy us with nuclear weapons. We can be certain they will retain that capability throughout the 1990's and beyond and we must deal with that.

We can choose to maintain either a retaliatory force designed to deter a Soviet attack, or we can maintain a force capable of striking first. If we choose a retaliatory strategy we will require 1000 nuclear weapons that must survive a nuclear attack. Currently we have approximately 20 missile submarines at sea at all times. These 20 submarines are armed and ready to respond to a Soviet attack with more than 3000 nuclear warheads. This is three times the amount needed to respond to a Soviet attack.

If we elect to maintain a capability to launch a surprise attack in an attempt to destroy Soviet strategic nuclear forces, including ICBMs [intercontinental ballistic missiles], air fields, naval bases and command posts, we will require no more than 3200 to 3800 nuclear warheads. We now have the capability and are ready if ordered to launch an overwhelming attack against the U.S.S.R. with 12,000 nuclear weapons, hence there is no military requirement to build more nuclear warheads during the 1990's.

If an agreement is reached to reduce the strategic forces in the U.S. and U.S.S.R. by approximately 50% then it probably will be necessary to cancel some of the new ICBMs, bombers, and Trident submarine programs currently being funded by the Congress. Major cutbacks in our warhead production could also result. The Soviet Union is already making unilateral cuts in some of its strategic nuclear weapons.

Some U.S. analysts and military planners worry a good bit about the possibility or, more precisely, the probability of the proliferation of nuclear weapons in third world countries. Many nations now have the nuclear material and the capability to build nuclear weapons, and we at the Center for Defense Information estimate that 15-20 countries could each construct some nuclear weapons by the end of the 1990's. No one seems to know what to do about this development but it is a factor which cannot be ignored.

Perhaps it would be prudent for the President to call a meeting of the existing nuclear nations and discuss the problem in hopes of finding a way to deal with it. Or perhaps an international meeting called by the Congress to examine the problem would be in order. . . .

Smaller Forces

Obviously we cannot predict the future course of events with certainty, but neither can we base our military strategy and force structure for the 1990's only on conditions which prevailed in the past. Hitler is dead, the Cold War is over, the Warsaw Pact and NATO [North Atlantic Treaty Organization] will soon be irrelevant. We can pretend to ignore the changes which are occurring at the risk of becoming irrelevant ourselves.

After World War II we promptly reduced our forces from 12 million to 1 1/2 million. After Vietnam we cut our forces from 3 million to 2 million and there is every reason now to move toward a force of about 1.2 million total active persons which could be operated for $100 billion less per year than the current $300 billion military budget.

Militarily, a total force of 1.2 million persons can reasonably be justified for the 1990's. However, there are a number of factors which will help maintain much larger and more costly forces.

Foremost of these is the large number of civilians who are now working in the defense industries: 3 million today as opposed to the 2 million when President Ronald Reagan took office in 1981.

"Hitler is dead, the Cold War is over, the Warsaw Pact and NATO will soon be irrelevant."

Many of our major industries have come to depend on profits from military contracts and are intent upon keeping their influence over members of Congress to maintain and even increase the level of military spending. We have created a permanent wartime economy, as President Eisenhower feared we might, which may overpower the efforts of those in the Congress who might favor a smaller, leaner military establishment.

Economic Needs

There is no need to spend lots of money on a huge "defense industrial base." It is extremely unlikely that the U.S. will be engaged with the Soviet Union in a rerun of World War II. Any war with the Soviet Union will be a nuclear war and short. We will not have

months and years to step up the production of tanks and battleships. We do not need to spend money for jobs and profits to keep companies in business who might make problematic contributions to the nation's real defense.

It is important that we maintain a high level of scientific and technological capability, but the pursuit of civilian research should take priority over exotic military activities reminiscent of the Star Wars boondoggle. Let the military be the recipients of spin-offs from civilian R&D [research and development] rather than making our struggling industries dependent on wasteful military R&D spending which does little to strengthen U.S. capabilities in world trade. . . .

Not Yet?

The Cold War has been ending for many years, perhaps since the Cuban missile crisis. There have, of course, been many detours and setbacks. There may be more in the future. But it is unmistakable that we are in a new era of human history. The Soviet Union is changing in remarkable ways. The United States will also need to change and adapt.

It is time to get on with discarding our outmoded Cold War military strategy and moving rapidly toward a smaller force structure. Some will say "not yet" but when we spend $6 billion every week on the military establishment we are engaged in monumental waste. Americans want a strong, effective military and they are rightly losing patience.

Rear Admiral Gene R. La Rocque is director of the Center for Defense Information, a Washington, D.C. organization which favors an effective defense but opposes excessive military expenditures.

"I cannot rationalize a real reduction in spending for defense based upon force drawdowns announced by the Soviet Union."

viewpoint 7

The U.S. Should Not Cut Defense Spending

William J. Crowe Jr.

Our global military posture is designed to keep the peace, prevent small crises from becoming big ones and deter major hostilities. In practice, we strive to maintain healthy strategic and theater nuclear capabilities, field high quality forces to offset numerically superior adversaries, exploit total force planning with our allies and limit war reserve material to the time required by industrial mobilization. This posture is not without fiscal constraints and military risks. Today, for example, we are not able to keep all of our forces, active and Reserve, ready for combat on an everyday basis. Similarly, we are not positioned to reinforce and defend simultaneously Western Europe, Southwest Asia and the Pacific Far East. Instead, we must plan to deal with these theaters sequentially, depending upon circumstances and priorities at the time.

Soviet Military Power

Meanwhile, we are confronted with a rather confusing and incomplete picture of the Soviet Union. In this situation, I am inclined to look at the essential elements and underlying pillars of Soviet military power—just as General Secretary Mikhail Gorbachev is doing today.

The basic characteristics of Soviet military power are well-known: robust capabilities in space, the world's only ABM [anti-ballistic missile] system, strategic forces with intercontinental range, conventional ground and air forces capable of power projection across the entire Eurasian continent, naval forces poised for the denial of Western sea lanes and armed surrogates near many of the world's oil fields and maritime choke points. Any U.S. military strategy must confront these realities.

The Soviet Union also towers above all other nations in the production of military equipment.

Reprinted from William J. Crowe Jr., "Cuts in the Defense Budget," *Defense 89*, July/August 1989.

Quantity is not the only worrisome part of this picture. For the last two decades, the USSR has been steadily closing the gap (relative to the West) in military technologies.

For example, the Soviet Union is on a qualitative par with the United States in such deployed systems as surface-to-air missiles, anti-tank guided missiles, tactical ballistic missiles, short-range naval cruise missiles, communications and electronic countermeasures. They hold an edge in anti-satellite systems, artillery, chemical weapons and mines. . . .

Strategic Nuclear Forces

With respect to the strategic nuclear balance, neither the United States nor the Soviet Union has neglected investments in strategic warning systems, and both have a credible capability to detect an all-out nuclear strike, although our warning system is more sophisticated.

Each side does approach strategic nuclear parity in a different way. For its part, the Soviet Union has more balance in strategic offensive and defensive systems. Their triad of land-based missiles, ballistic missile submarines and bombers is complemented by substantial investments in leadership survival, strategic air defense, anti-satellite devices and the world's only ABM system. In the Krasnoyarsk facility, they have indicated a willingness to go beyond limits of the ABM treaty.

Additionally, Soviet strategic offensive forces are more dispersed and hardened than those of the United States. In the case of ICBMs [intercontinental ballistic missiles], they have exploited government control of vast lands within the Soviet Union and placed most of their missiles in very hardened silos or on mobile launchers.

When it comes to military targets, our computer-supported analyses always show that Soviet ICBMs give the Kremlin a relative advantage in "damage expectancy." This lead, however, is only one of several factors bearing on the nuclear calculus.

For its part, the United States must maintain general nuclear parity with the Soviet Union by means of strategic modernization programs set in motion over the last decade: more firepower in the Trident-class submarine, improved accuracy and hard-target kill capability in the submarine-launched D-5 missile and the land-based Peacekeeper missile, technologies seen in the B-2 and the long range and high accuracy of air-launched cruise missiles.

Dependence on Nuclear Forces

These strategic modernization programs are "vital" as the Soviets continue to improve their active and passive defenses and, especially, as we move into a Strategic Arms Reduction Talks regime. Put bluntly, a healthy strategic deterrent is the cornerstone of our overall defense posture and compensates to a real extent for some of the shortcomings in our theater nuclear and conventional forces. It serves a similar function for many of our allies.

Further, research on the Strategic Defense Initiative tends to discourage a larger Soviet investment in offensive ballistic missiles and serves notice that we will not allow Moscow to monopolize the field of ABM systems. We are achieving impressive, tangible results with this program as we examine cost-effective ways to move toward a safer defense posture.

Starting in the mid-1960s, NATO [North Atlantic Treaty Organization] became increasingly dependent upon theater nuclear forces to compensate for conventional shortcomings and provide options short of a strategic nuclear exchange. The USSR responded in kind and eventually outnumbered the United States in intermediate- and shorter-range missiles. The Intermediate-Range Nuclear Forces Treaty will eliminate these types.

Major asymmetries continue in short-range nuclear missiles, favoring the USSR. Additionally, NATO's practice of relying upon dual-capable tactical aircraft raises worrisome questions about the survivability and attrition of these assets at the outset of a conventional war. In part, we can hedge this risk by relying more on rear-area bases or seaborne platforms, but the supreme allied commander Europe urgently needs the tactical air-to-surface missile and a modern nuclear battlefield system as seen in the follow-on to Lance. . . .

The Current Prospects

Today, we are looking at the prospect of a net decline in U.S. military capabilities. This trend is related directly to the federal budget deficits and a reversal of the earlier defense buildup.

From fiscal 1985 through fiscal 1989, real spending for defense (adjusted for inflation) declined by 12 percent. In turn, about $300 billion was cut from the defense plan contemplated in January 1987. . . .

Put bluntly, as the senior adviser to national command authorities, I would vastly prefer a dollar figure that would permit us to keep our current force structure (without sacrificing quality) until we have a clearer understanding of where the Soviet Union is going, of the arms reduction calculus and of the international climate. In my judgment, there are too many uncertainties on the horizon at this time to justify force cuts.

America will enter the 1990s spending little more than 5 percent of its gross national product on defense. This is a very modest investment—perhaps too modest—considering the size of our gross national product, international conditions, the military threat, our global interest and the forces needed to support a longstanding policy of "peace through strength."

Recent military initiatives by General Secretary Gorbachev are encouraging. Yet, I find more continuities than changes in the essential elements of Soviet military power. Moreover, we have yet to see a larger commitment of Russian resources to consumer goods and services. Until this happens, the future of *perestroika* and *glasnost* remains cloudy.

"Major asymmetries continue in short-ranged nuclear missiles, favoring the USSR."

Between 1985 and 1989, the Department of Defense absorbed an 11 percent decline in real spending for defense and still managed to maintain a reasonable balance between its global responsibilities. On NATO's Central Front, however, the military risks are higher than I and the Chiefs would prefer. Today, we are challenged to pursue such military objectives as a credible nuclear deterrent, a secure position in space, a healthy maritime environment and strategically mobile ground and air forces. Concurrently, our military establishment is more operationally involved in the national drug-interdiction effort.

In essence, I cannot rationalize a real reduction in spending for defense based upon force drawdowns announced by the Soviet Union, shrinking demands upon our forces or the prospect that an arms reduction agreement is just around the corner.

Retired Admiral William J. Crowe Jr. was the chairman of the Joint Chiefs of Staff from 1985 to 1989. In addition, he was formerly commander in chief of the Allied Forces in Southern Europe.

"[Defense] cuts . . . appear certain to bring an eventual 'peace dividend' to the U.S. in the form of lower inflation and interest rates, a declining budget deficit, and faster growth."

Cutting Defense Spending Will Help the U.S. Economy*

Karen Pennar and Michael J. Mandel

*The title of this viewpoint is "The Peace Economy." See footnote below.

There's no set agenda. No agreements, hammered out over months, waiting to be initialed. But the whole world is watching nonetheless.

As George Bush and Mikhail Gorbachev meet aboard the USS Belknap and the Slava off the Mediterranean island of Malta, they will be opening a new chapter in East-West relations. Across Eastern Europe, cries for freedom have toppled walls—and governments. The Soviet leader's call for *glasnost* and *perestroika* at home has unleashed a whirlwind of change abroad. These days, anything can happen —and usually does.

Now, the superpowers have an opportunity to reinforce change. They are already winding down defense spending in the wake of intermediate-range arms control negotiations and easing tensions. But the developments in Eastern Europe are sure to accelerate cutbacks: Against a backdrop of rapid liberalization, the Warsaw Pact threat seems to be diminishing by the day, and NATO's [North Atlantic Treaty Organization] role will perforce change. The geopolitical assumptions of nearly half a century of cold war have been undermined, and an era of disarmament may have begun.

But it's not peace so much as economics that's driving military retrenchment—and Gorbachev was the first leader to embrace the economic logic of defense cutbacks. The U.S. and the Soviet Union have spent years battling for military supremacy, only to see Japan—which spends a mere 1% of its output on defense—run away with the world's economic prize. Gorbachev, struggling against economic stagnation at home, desperately needs to hack away at the 16% of output devoted to the Soviet military. And Bush, facing a budget gap that won't close and fierce economic rivalry with Japan, could certainly use defense savings.

"It's good for them and it's good for us," says Nobel laureate Paul A. Samuelson. "In any well-run society, it's still a question of guns or butter. Only in times of tremendous unemployment can both be afforded."

Making a Dent

By the mid- to late-1990s, defense experts and even some contractors are saying military spending could decline from about 6% of U.S. gross national product to around 4%. That would be the lowest level since the massive demobilization following World War II, and the closest this nation has come to a peace economy in the postwar years. And cuts in spending could turn out to be deeper than those now being contemplated in Washington.

The peace economy could look very different. Spending cuts of the magnitude that the Pentagon is examining would, at the very least, put a dent in the intractable $150 billion budget deficit—and pull down interest rates and inflation. That, in turn, should enhance investment and boost housing activity. Part of the budget savings, meanwhile, would almost certainly be diverted to spending on infrastructure and education, which could help to enhance lagging U.S. productivity. By the end of the century, U.S. GNP could be growing nearly 20% faster than it would be without the cutbacks, according to calculations made by economic forecaster DRI/McGraw-Hill for *Business Week*, based on the cuts now being discussed.

This military retrenchment marks a dramatic reversal of the Reagan buildup of the 1980s. Strategic thinkers will argue endlessly over whether Reagan's defense push was a necessary condition for bringing the Soviets to the bargaining table. If it was, then those policies may have been worth their cost. But years of nuclear proliferation, conventional-arms buildup, and battles by proxy around the globe have exacted a stiff price from both nations.

Military spending does create jobs: Each $1 billion

cut in Pentagon outlays affects 38,000 U.S. workers, according to Employment Research Associates in Lansing, Mich. But that simple calculus obscures the deadweight cost of more arms. Long-term growth and prosperity are hampered as technical talent and research dollars are diverted to the military. Massive defense outlays aggravate the budget deficit. And there's the opportunity cost of pouring money into armaments rather than education, infrastructure, and the environment.

"If . . . 5% annual cuts were adopted, by the mid-1990s the amount spent on the military would about equal consumer spending on clothing."

No one knows how deep or how far-ranging cutbacks in military spending are likely to be. Defense Secretary Richard B. Cheney has instructed the services to look at cuts of up to $180 billion for fiscal years 1992-94. That figure, it turns out, represents cuts from projected increases in the defense budget. Still, after accounting for inflation, what Cheney has proposed amounts to cuts, in real terms, of up to 5% a year. If the reduction in fiscal 1991 is about the same size, as many analysts expect, then the spending rollback from 1991 to 1994 would total about $60 billion in 1989 dollars, for a 20% reduction in real terms. Barring a sudden reversal in the course of East-West relations, that seems a reasonable course of spending to expect.

But some defense experts are already arguing that the cuts could go far deeper. Brookings Institution senior fellow William W. Kaufmann provides a blueprint for steady and deep cuts throughout the 1990s, which would shrink the Pentagon budget by up to 50% in real terms by 1999. Kaufmann, a former Defense Dept. adviser, says that such a build-down could be halted and reversed, if necessary. Kaufmann would hold big-ticket items such as the $45 billion Advanced Tactical Fighter or the $35 billion Advanced Tactical Aircraft in the research and development and testing phases. He also proposes holding back on new land-based nuclear weapons and cutting the Air Force's $70 billion plan for the B-2 or Stealth bomber to 13 from the current 132.

Peace Euphoria

Shelving big programs is generally viewed as a good budget-paring strategy. For now, though, it's personnel that are likely to be the first target for cuts. As Bush headed for Malta, the proposals for conventional arms reduction introduced in May 1989—putting U. S. and Soviet troops in Europe on an equal footing at 275,000 each—were already obsolete, thanks to events in Eastern Europe. To get to that level, the U. S. would

pull out 30,000 troops, while the Soviets would withdraw about 10 times that number. For the U.S., savings from further cutbacks could be substantial: As much as 60% of the defense budget represents the direct and indirect costs of defending Europe.

Despite the current peace euphoria, winning congressional approval of large cuts won't be easy. For one thing, there's the argument in favor of prudence: Surely it's better for the U.S. to wait and see how successful Gorbachev is at turning Soviet opinion and the Soviet economy around. If Gorbachev himself is unseated, then U.S. military cutbacks might look foolhardy in retrospect. More fundamentally, however, many in Congress view defense contracts as jobs programs—gifts they can promise their constituents and items they can trade support for with legislators from other districts or states.

Indeed, Congress exercises such control over defense budgets that even when the Pentagon doesn't want some hardware, it still gets it. Cheney's attempt to ax the V-22 Osprey tilt-rotor aircraft and the Grumman F-14D fighter met with stiff resistance on the Hill, and proposed base closings have also been hotly contested. The next round of debates should be no smoother. Says a Washington lobbyist for a major defense contractor: "Congress has become accustomed over the years to micromanaging a growing defense budget, and now we can't ask them suddenly to take their hands off."

Still, it looks as if both Democrats and Republicans will be eyeing rollbacks in spending that would have been unthinkable only a few months ago. And while many defense contractors doubt the cuts will be deep, some are preparing for a sharp retrenchment. John D. Rittenhouse, senior vice-president of General Electric Aerospace Group, says that for more than a year he's been expecting real cuts of 5% to 10%. "There is no news here to defense contractors who were attentive," he says of Cheney's proposed cuts.

Lower Inflation

Whatever the size of the cuts, they appear certain to bring an eventual "peace dividend" to the U.S. in the form of lower inflation and interest rates, a declining budget deficit, and faster growth.

In an attempt to measure the impact, *Business Week* asked DRI/McGraw-Hill to look at the results of cutting defense by 5% a year in real terms for four years starting in 1991, with no growth thereafter. This would yield a cumulative cut in defense outlays of approximately $60 billion in 1989 dollars, close to the upper limit of what the Cheney cuts would achieve by 1994. DRI compared the results with its long-term "base case," which assumes more modest declines in defense spending through 1992, followed by average annual real increases of 1.8%.

If the 5% annual cuts were adopted, by the mid-1990s the amount spent on the military would about equal consumer spending on clothing. The defense budget would shrink from about one-quarter of total government spending to about one-fifth. The budget gap would be halved by 1995 and continue moving toward balance. With the deficit shrinking, interest rates would plummet. The Federal funds rate would fall to 5.5%.

Indeed, because this regimen has such salutary effects on the budget deficit, DRI assumed that when the budget turned to surplus in 1998, that surplus would go toward additional nonmilitary government spending. This diversion of savings from defense might occur even sooner. Whenever it does, the additional spending would give the economy a boost above and beyond the effects of low interest rates. The end result: much higher rates of growth.

Under the scenario of 5% cuts coupled with extra nonmilitary spending at the end of the decade, DRI found that economic growth, after slipping below DRI's long-term base forecast early in the decade, would start growing faster than the base case in 1995. By 1999, the real gross national product would be rising at a 2.6% real rate, compared with 2.1% under the base case.

"For companies that prepare themselves, there is life after defense contracting."

The DRI results show that the defense cuts, and the lower interest rates they bring, should energize U.S. investment. From 1995 on, the economy would add plants and equipment at a faster rate. And the international competitiveness of U.S. companies would be enhanced. Right now, they are struggling against Japanese and West German competitors that face much lower capital costs. According to a Federal Reserve Bank of New York report, foreign companies pay an inflation-adjusted, aftertax cost of funds as low as 2%, compared with 6% in the U.S. Lower interest rates in the U.S. will help narrow the difference.

Among the other payoffs: Housing starts would start rising by 1992, and the peace economy would produce 500,000 more housing units by the year 2000. Demand for autos and consumer appliances would be stronger, too. The machine-tool industry, not heavily dependent on military contracts, would come out ahead. The trade gap would shrink, since the federal government would have to borrow less from foreigners. And the Standard & Poor's 500-stock index would be 15% higher by the end of the decade than if defense weren't cut.

However, the road to the peace economy may be bumpy. With defense contracts concentrated in a relatively small number of industries and regions, the localized hits would be dramatic. Two-thirds of the R&D [research and development] contracts and 40% of procurement contracts are concentrated in 15

metropolitan areas, with Los Angeles the biggest. The most vulnerable industries are aerospace and ordnance. But 34% of the business for optical-instruments makers comes from military contracts. And 63% of the domestic radio and TV equipment industry's output goes to the military.

No Worries?

For now, the prospect of the Cheney cuts doesn't have many contractors or state officials troubled, and deeper cuts don't seem to be a concern. Even executives at those contractors most heavily dependent on the defense dollar, such as McDonnell Douglas Corp. and General Dynamics Corp., seem sanguine. William S. Ross, president of McDonnell Aircraft Co., doesn't believe that cuts will go as deep as Cheney has suggested. And GD President Herbert F. Rogers believes the company can weather individual program hits because it's involved in programs that cut across the armed services. "We'll shrink or grow as the defense budget shrinks or grows," says Rogers. "It's not axiomatic that you won't do well [if budgets shrink]."

No matter how individual companies respond, the industry's ongoing shakeout will only be accelerated. According to a study by the Center for Strategic & International Studies in Washington, there were more than 138,000 companies providing manufactured goods to the Pentagon in 1982. Five years later, there were fewer than 40,000. In part a "grossly inefficient" defense-acquisition system was turning off suppliers, the study found. Further, defense hasn't been steadily profitable. Return on sales from defense manufacturing fell from 4.9% in 1980 to 3.8% in 1986.

For companies that prepare themselves, there is life after defense contracting. In 1986, government defense work accounted for 50% of Rockwell's sales. Today, that figure is down to 28% and it should fall to 25%. Donald R. Beall, chief executive at Rockwell International Corp., says the company began planning in 1985 for the end of the B-1 bomber program in 1988. Rockwell made some acquisitions that are now paying off and began pushing nondefense electronics, computerized manufacturing, and civilian space work.

Seattle's Boeing Co., thanks to booming demand for its commercial jets, has seen the mix of its business change markedly: By 1991, analysts expect Boeing's commercial sales to represent 75% of its business, up from 62% in 1987. For Boeing, defense has become a money-losing business anyway. And, like Boeing, helicopter maker Sikorsky is benefiting from a pickup in foreign demand. Today, Sikorsky gets 25% of its revenues from exports, up from a scant 6% in 1984.

Double Whammy

But while individual companies may be able to protect themselves, regions could have a harder time of it. In Massachusetts, for instance, defense revenues have dropped 17% from 1986 to 1988. New England could lose 11,000 defense-related jobs in 1990, according to DRI, with 60% of those in the Bay State. The defense slowdown coming on top of the computer slump will be a double whammy for the state.

Even Southern California, with its highly diversified economy, won't be able to avoid taking a hit. "The local economy," says David Hensley, of the UCLA [University of California, Los Angeles] Business Forecasting Project, "is definitely going to feel this. What it won't do is push us into a regional recession." That's because of extensive diversification in recent years. At the peak of military spending during the Vietnam War, aircraft and missile jobs accounted for 4.4% of total employment in the state. But in 1987, the peak year for defense outlays during the Reagan buildup, those jobs amounted to 2.2% of the state's work force.

Orange County officials are meeting with local contractors to figure out ways to cushion the blow from possible cutbacks by such local employers as Northrop. "We don't want to see people lose their jobs or miss out on the scientific genius now being applied to the military," says Larry Agran, mayor of Irvine. And in Ohio, which has some 200,000 people working on defense contracts and ranks eighth in the nation in prime contracts, Governor Richard F. Celeste is bringing small business contractors together to discuss adjusting to a shrinking defense budget.

Base closings offer communities some difficult choices, though there have been some notable successes. Stewart Air Base outside Newburgh, N.Y., has been partially converted to civilian use and now functions as Federal Express Corp.'s East Coast hub. And American Airlines Inc. will be starting passenger service out of Stewart. Bases such as San Francisco's Presidio present no end of possibilities for conversion. The base, perched on a promontory overlooking the Golden Gate Bridge, represents 1,440 acres of choice, forested real estate. Of the 800 structures on the base, 400 are protected as historical sites. Planning is just getting under way to determine the Presidio's future. Turning it into an educational center or a Pacific Rim conference site are two possibilities actively being considered.

Adaptability

Local officials only recently began preparing for the time when military cutbacks shrink local payrolls, and some say the federal government will have to help out. Two economic conversion bills are under consideration in Congress, offering ways to aid communities, including money for planning, special unemployment assistance, and retraining. House Majority Leader Richard A. Gephardt (D-Mo.) supports conversion legislation, and Senate Majority Leader George J. Mitchell (D-Me.) intends to promote the concept. Without such planning, says Seymour Melman, a member of the National Commission for Economic Conversion & Disarmament, in

Washington, there will be widespread economic distress when some regions lose defense contracts.

The real issue for companies and, ultimately, for communities, is adaptability. Says John M. Kucharski, chairman of EG&G Inc. in Wellesley, Mass.: "If the defense budget does take a dramatic cut, will the money stay in the public sector? I believe it will." And in that case, the federal government will spend more on infrastructure and the environment, providing an opportunity for companies nimble enough to switch gears.

"The real issue for companies and, ultimately, for communities, is adaptability."

There's no lack of demand for spending on roads, bridges, and airports. According to a study from the Congressional Budget Office, the U.S. falls about $15 billion short of its spending needs on transportation, infrastructure, and water treatment each year. Other estimates of the nation's annual infrastructure deficit run as high as $100 billion.

But even if the U.S. was willing to invest only $15 billion more each year on its infrastructure, it could show big payoffs. David A. Aschauer of the Federal Reserve Bank of Chicago estimates that this level of increased spending on core infrastructure—utilities, roads, rails, and airports—could boost the productivity growth rate by about 0.3 percentage points.

There's no telling, of course, just how the savings from defense cutbacks will be allocated in coming years. For the Democrats, the new moneys represent an opportunity to divert resources toward other human capital uses. For Republicans, the funds represent a way to reduce the deficit and help avoid a tax hike. The battle over these defense savings—how large they'll be and where they'll go—is certain to rage for years to come. But America's leaders have an opportunity to demonstrate that for the U.S. and the rest of the world, the politics of peace can be the politics of prosperity.

Karen Pennar is the economics editor of Business Week. *Michael J. Mandel is associate economics editor for the magazine.*

Cutting Defense Spending Will Not Affect the U.S. Economy

Murray Weidenbaum

The perennial debate in the United States on the impact of defense spending on the economy has been heating up. Those who favor smaller military budgets cite the high "opportunity cost" of diverting vital scientific and technological resources from productive civilian pursuits—a diversion, they argue, that undermines productivity at home and competitiveness abroad. Advocates of this view also try to show that a dollar (or rather a billion) for defense produces fewer jobs than the same amount of money devoted to non-military expenditures. High levels of defense spending, they conclude, are economically unsound and sap the nation's prospects for growth.

Another widely circulated criticism is that of historian and best-selling author Paul Kennedy, whose *The Rise and Fall of the Great Powers* warns that too large a proportion of a nation's resources allocated to military purposes rather than "wealth creation" is likely to lead to "a weakening of national power over the longer term." Kennedy specifically raises the specter of "global overstretch" on the part of the United States. (On occasion, he refers to his concern as "imperial overstretch.")

Advantages of Spending

For their part, the proponents of larger military budgets cite what they believe to be the signal advantages of defense spending. Chief among these, they argue, are the favorable "spinoffs" of defense technology into high growth electronics, instruments, and aerospace industries. These advocates also focus on the large number of high-paying industrial jobs created by military outlays (35,000 jobs for each $1 billion of defense spending, according to former Secretary of Defense Caspar Weinberger).

It is fascinating to compare these two sets of self-serving arguments, for they are literally mirror images of each other. Both camps are united by the idea that defense spending has powerful impacts on the economy, whether for good or ill. Yet the truth seems to be quite different. The economic experience of the period since World War II shows that both critics and supporters of defense spending have seriously overestimated the importance of what they have found.

Defense spending generates some broader benefits, but the costs are substantial. Military research and development does produce important technological "spill-overs" into the civilian sector. The education, training, and physical conditioning that young men and women obtain in the armed forces are of obvious benefit to society as well as to themselves—especially when those skills are applied to civilian occupations. However, military outlays are rarely the most efficient way of securing these desirable side effects. A new treatment for AIDS, to take one example of obvious importance, is more likely to come from medical research than from work on the strategic defense system.

The naysayers on defense have likewise overstated their case. Despite high levels of defense spending, new civilian jobs are being created rapidly in the United States—far more rapidly than the nations in Western Europe who devote much smaller shares of their GNP [gross national product] to defense. Since the end of World War II, in fact, the relative importance of defense to the economy of the United States has been declining. Different ways of gauging resource use over the past half century yield the same point: defense outlays have accounted for a declining share of the GNP; defense spending has been a declining portion of the federal budget; defense manpower has represented a declining fraction of the nation's work force; defense has received a diminishing portion of the nation's research and development funding.

Military outlays now represent only one-fifteenth of

Murray Weidenbaum, "Why Defense Spending Doesn't Matter," *The National Interest*, no. 16, Summer 1989. © 1989 *The National Interest*, Washington, D.C. Used with permission.

the GNP and an even smaller proportion of the nation's work force. Few of the largest industries produce significant portions of their output for the military. Many of the major defense contractors sell the bulk of their products in civilian markets. Moreover, long periods of relative decline in the military's use of research and development and other high-powered resources have not resulted in comparable increases in civilian demand, much less a pick-up in U.S. productivity and growth rates. The decade from the mid-1960s to mid-1970s provides a striking case in point.

More recently, an analysis of the economic impact of the military buildup of the early 1980s found no evidence of any "major disruptive effect" of defense expenditures. No substantial bottlenecks were encountered. If anything, defense spending served as an unplanned counter-recessionary force in the 1981-82 downturn.

The economy of the United States is both complex and massive. It is not readily propelled or retarded by the relatively small share of GNP devoted to military outlays, and its powers of adjustment are substantial.

What then has been the *actual* impact of military spending on the United States and its position in the world? From a modest level of about $1 billion in fiscal year 1938, the outlays of the Department of Defense rose to approximately $285 billion in the fiscal year 1988. That, of course, was a far more rapid increase than occurred in the population of the country or the size of the economy, or both. A similar upward trend is visible if the data are corrected for inflation—or if manpower levels are used instead of dollar figures, although the annual fluctuations are quite different in some time periods. Because the overall American economy was expanding during the same period, it is useful to focus on the changing relative position of military outlays.

The Post-World War II Pattern

The most substantial absolute and relative expansion in U.S. defense expenditures occurred during World War II. In the years since the deep and rapid postwar demobilization, two limited-war expansions occurred (Korea and Vietnam), plus a buildup in the early and middle 1980s.

The most important fact that emerges from the historical record is that the relative importance of defense to the American economy has been declining since the end of World War II. To be sure, the pattern is uneven. Nevertheless, the Korean peak of 14 percent of GNP was far below the World War II high of 39 percent, and Vietnam War outlays were proportionately lower (less than 10 percent of GNP) than during the Korean period.

The high reached in the Reagan administration was a comparatively modest 6.5 percent in 1986 and 1987—a ratio that was exceeded in many peacetime years in the 1950s and 1960s. Declines in that ratio

are almost inevitable in the near future because of the substantial reductions in defense appropriations, and hence in the ability to make forward commitments, that Congress has enacted during the last several years. Thus, over a very significant time period, military activities have been a steadily smaller factor in the American economy. In striking contrast, civilian spending has been the growth area of this nation's economic activity. In the 30-year period from 1958 to 1988, the military share of the U.S. GNP declined in 17 years, was stable in 1, and rose in 12 years.

The Changing Role of Defense

It is also instructive to evaluate the changing role of defense in national priorities. The most widely used measure of that relationship is to estimate the share of federal government outlays directed to defense. The trend here is basically similar to that for the GNP. The large portion of the federal budget directed to defense spending during World War II—over 90 percent—has not been equalled since. A secondary peak occurred during the Korean War, when defense spending accounted for almost 70 percent of the budget.

"The economic impact of defense activities in the United States peaked decades ago and has been declining—albeit irregularly—ever since."

Since then, the defense share of the federal budget has declined to a low of less than 23 percent in 1980. It reversed to a high of 28 percent in 1987 and is now declining. Account should be taken of the substantial expansion in the scope of federal civilian responsibilities during the 1960s, especially the Great Society programs. To a large extent, therefore, the decline in the military share of the federal budget resulted from the fact that civilian program outlays were growing at a more rapid rate.

It is ironic to note that Caspar Weinberger's staunchness in preserving the defense budget in the early 1980s made it politically difficult to make deep cuts in non-military expenditures. Proponents of the latter raised the issue of "fairness" in limiting reductions in non-defense programs. During the 1980s, federal civilian expenditures continued to rise in real terms and also tended to maintain a relatively constant share of the GNP. This experience runs counter to the common belief that expansions in defense spending invariably come at the expense of civilian government outlays.

Nevertheless, military outlays are now under considerable pressure because of general budgetary trends in recent years. A combination of rapid expansions in both military and civilian spending programs in the early 1980s, coupled with substantial reductions in income tax rates, led to unparalleled large budget deficits. The persistence of these triple-digit deficits beyond the 1981-82 recession led to institutional restraints on federal spending (the Gramm-Rudman-Hollings legislation). The military budget was a major target of Gramm-Rudman-Hollings and appropriations for the Department of Defense by the mid-1980s were less than the amount necessary to keep up with inflation.

What should we make of the concern over the American military's supposed ill-use of key resources? An analysis of the changing importance of the military demand for key factors of production is revealing; it hardly supports the contention of a society "depleted" by a military establishment hogging the vital resources of the nation. Thus, the armed forces now represent only 1.7 percent of the total U.S. labor force, down from a peak of 4.3 percent in 1955, but also down from 2.2 percent in 1975.

Decreasing Economic Impact

During the same general period, the proportional decline in the military share of funding for research and development has been dramatic. In 1960, the Department of Defense obtained the lion's share of the nation's scientific and technological resources—62 percent. By 1980, the ratio had plummeted to 25 percent. In 1987, the preliminary data show a 30 percent share for the military—less than one-half the 1960 proportion.

Surely, the absolute size of defense purchases of goods and services looms large for all the available statistical measures. The Department of Defense is a major "customer" of American business. Nevertheless, the overall pattern is clear: the economic impact of defense activities in the United States peaked decades ago and has been declining—albeit irregularly—ever since.

"It has become fashionable to equate the comparatively large percentage of U.S. GNP devoted to defense with the slippage in the U.S. share of world trade."

Whatever their relative or absolute size, the resources allocated to national defense are not available for civilian purposes. Especially in a fully employed economy, it is reasonable to assume that, in the absence of the military's demand, much of those resources would have gone to meet civilian needs. The question then arises as to which areas of the civilian economy would use the resources that would become available following a reduction in military budgets. To an economist, the "opportunity cost" of expenditures for defense is the opportunity forgone to use the people, machinery, and materials in some other ways.

Do increases in defense spending come primarily out of resources that otherwise would be devoted to investment (a primary ingredient in economic growth)? To the extent that such is the case, the "opportunity cost" of defense spending is high. Every dollar devoted to defense would mean a dollar less invested in the future of the economy.

No Causal Relationship

On the other hand, if the money spent for defense would otherwise go for current consumption—for items that generate little or no future benefit—then the true cost of defense is transitory and much lower. Researchers who have looked into the matter have found less than universal agreement. The possibility of defense demands crowding out private investment rests on the notion that a large and growing federal deficit forces the Treasury to expand its presence in capital markets, putting upward pressure on interest rates. Rising interest rates, in turn, inhibit capital formation. However, the empirical evidence on the causal relationship between budget deficits and interest rates is not very impressive.

It turns out that, in most cases, increases in the share of GNP devoted to defense are accompanied by reductions in the proportion going to consumption, rather than to investment. Kenneth Boulding has obtained such results using an analytical approach based on national income accounts. The substantial rise in personal income tax collections in the period since World War II helps to explain this trend.

Another charge one often hears is that military spending on research and development "crowds out" civilian R&D [research and development]. Let us examine the period between 1949 and 1988, for which detailed data are available. In only 16 of those 39 years did the military and civilian shares of the federal budget move in opposite directions. In 18 of those years, the shares of the federal budget devoted to civilian R&D and to military R&D moved in the same direction—the civilian R&D portion rising when the military R&D share rose, and falling when the military share fell. In 5 other years, the civil sector registered no change in its share of the federal budget.

The "depletion" thesis does not hold up. Expanding military R&D is at least as likely to have a positive effect on civilian R&D as the negative impact that is so often envisaged. Moreover, reducing the military R&D share of the federal budget is as likely to have a negative effect on civilian R&D as a positive effect.

There is no shortage of studies that purport to show an inverse relationship between the concentration of a nation's economy on defense and its poor economic performance. Thus, the argument goes, the United States spends proportionately more on defense than Japan and therefore we have a consistently lower rate of economic growth. Yet South Korea, which devotes a larger share of its GNP to defense than Japan, boasts a more rapid growth rate. To jump to a heroic conclusion from either comparison is surely simple-minded.

Other factors—such as the national saving rate—are important influences on a nation's growth rate. Still, since it has become fashionable to equate the comparatively large percentage of U.S. GNP devoted to defense with the slippage in the U.S. share of world trade and global economic activity, let us pursue that point.

Losing Economic Power

It is easy to show that the United States has lost its supremacy in the global economy in the four decades since the end of World War II. In 1950, the gross national product of the United States represented approximately 45 percent of the world's gross product. In the last few years, in striking contrast, the U.S. share has dropped to about one-fourth of the global total.

Again, some historical perspective is useful. In 1950, the economies of Western Europe and Japan were still recovering from the devastation of World War II. Under those circumstances, the American economic giant had little difficulty dominating world markets. But such a powerful position was bound to be transitory, as the economic competitors regained their traditional strength—with much help from the United States. The current relative position of the United States is little different from what it was in 1938.

It is intriguing to note that the Soviet Union did not take such a benign attitude. It shackled the economies of defeated nations within the sphere of its control. The poor economic performance of the Soviet bloc economies in the period since World War II, however, is hardly a tribute to that approach.

Statistical comparisons, favorable or unfavorable, have their limitations. Thus, in the 1950s and 1960s—when the economic power of the United States was rarely questioned—a rapid spread of collectivist and anti-market policies occurred in many parts of Western Europe and Asia. In the 1980s, however, during the period of supposed U.S. decline, this trend has been reversed. In many parts of the world a dramatic expansion has occurred in the role of market forces, economic incentives, price competition, and the privatization of economic activity. Great Britain and China provide two very different but equally impressive examples of this powerful change.

Doom Peddlers

The doom peddlers always seem to have a field day in competing for public attention. Yet, the United States remains the leading power in the world. In 1988, America's farms, mines, factories, and offices produced almost $5 trillion of goods and services—a

record high and more than double second-place Japan's GNP of just over $2 trillion.

Any realistic assessment of the position of the United States in the world economy must take account of important new factors in the economic equation. To a substantial degree, the impact of domestic considerations such as defense spending is overshadowed by the new competition from an array of developing countries that have joined, or are about to join, the club of advanced industrial nations.

Economic history provides a useful perspective. In the nineteenth century, European investors financed much of the canals, railroads, and heavy industry that enabled the United States to become a global economic power. But that also eliminated the European monopoly over the world economy. Nevertheless, Europe's international trade continued to rise substantially in absolute terms. Something similar is underway today. Investment funds provided by the United States and the other developed nations have helped to create a new set of actors on the world economic stage, mainly in the Asian rim. Once again, the return to the status quo ante is not in the cards. In the short run, the adjustments are painful to many established sectors of the more advanced societies.

"The U.S. position in future international rankings will depend in large measure on matters quite independent of the military."

Over the long run, these changes in the economic power of individual nations make for a stronger international commercial system. Our best customers today are the other advanced economies. The resulting expanded flow of international trade and investment yields higher living standards for consumers in general. That was the experience of the nineteenth century, and it is being repeated as we approach the twenty-first century.

Looking to the Future

None of this should be taken to minimize the importance of fiscal prudence. Of course the portion of our national resources devoted to military purposes should be carefully scrutinized; of course the serious shortcomings in the military procurement process should be dealt with promptly.

But the U.S. position in future international rankings will depend in large measure on matters quite independent of the military. These include controlling production costs, enhancing productivity, improving the education of our work force, and promoting national competitiveness in other ways. The outcome, given some tough decisions on public budgets and private productivity, is not likely to be as dismal as the doom peddlers would have us believe.

One experienced observer, Zbigniew Brzezinski, predicts that in the year 2010 the United States and the European Economic Community will be the two dominant forces in the world economy. In "America's New Geostrategy," an article that appeared in *Foreign Affairs*, he estimated that each will generate in that time period an annual GNP of approximately $8 trillion—double or more than that of Japan, China, or Russia. That result would not be too shabby for a nation so heavily criticized for "overstretch." If anything, it is the Soviet Union—which both devotes a far larger share of its national resources to military purposes and suffers from a combination of low productivity, slow growth, and great pressures of unmet civilian needs—that should be concerned about "overstretch," to say nothing of "imperial" overstretch.

A Political Issue

In sum, the U.S. military budget could vary over a considerable range without raising the specter of economic harm or national decline. This is not a plea for adopting the high end of that range, or for otherwise assaying the desirable size of the military budget. But the amount of resources that the United States devotes to defense programs should be determined primarily on non-economic, and essentially political—that is, national security—grounds, with little fear of undermining this nation's position in the world.

Murray Weidenbaum is Distinguished Scholar at the Center for Strategic and International Studies, a Washington, D.C. research center which studies international business and economics as well as arms control. He is the former chairman of the president's Council of Economic Advisors.

"The main threats to our international position are domestic. . . . This is the realm in which we must establish our strength."

viewpoint 10

Defense Spending Should Be Cut to Fund Social Programs

Jack Beatty

The polls show that growing numbers of Americans want this country to respond to the world-historical developments of what deserves to be called the Gorbachev era with something that matches them in scope. It's hard to see what that could be in foreign policy—after all, we have no Berlin Wall to tear down. To be sure, a strategic-arms-reduction treaty and a treaty reducing conventional arms in Europe would be signal achievements, and maybe if both treaties had already been signed, they would satisfy our wish to make the most of this historic moment. But perhaps looking for ways to match Mikhail Gorbachev in foreign policy is missing the point of Gorbachev. As Paul Kennedy has argued, Gorbachev is the only practitioner of "grand strategy" on the contemporary world stage, and the essence of his strategy is a manifest willingness to subordinate foreign and defense policy to the domestic necessity of making the Soviet economy work. Just when what Kennedy calls "the metric of power" in world politics has begun to shift from military force to economic potential, the Soviet Union has been graced with a leader who appears to understand that shift, and who is embarked on a desperate gamble to restore the greatness of his country. Gorbachev's grand strategy makes possible the kind of change in our own national strategy that would have been unthinkable at any other point since the Cold War began.

The United States today faces no external threat from a rising "challenger state." The main threats to our international position are domestic in kind; they are to be found in the debt-ridden condition of the economy and the deteriorating state of so much of our physical and human capital. This is the realm in which we must establish our strength, for it is here that we will be tested in the post-Cold War era. The metric of power points home, and it is this new idea of strength that the White House and Congress have the chance to codify. They have a choice, really, between Cold War and post-Cold War ideas of strength, between yesterday and tomorrow. What follows is an outline of some of the choices open to them—and us.

We could build the Stealth bomber. At nearly $600 million per plane, the Stealth is notoriously exorbitant. It represents the Air Force's attempt to keep the manned bomber alive in the age of the cruise missile, or pilotless flying bomb. As such, it is an exercise in nostalgia, one that could easily cost $79 billion—the "if all goes well" cost of the 132 planes requested by the Air Force. So far, Congress has spent $22 billion on the Stealth. What has all this money been for? It's dismayingly hard to tell. "The mission [of the Stealth] changes just about every week," says Representative John Kasich, an Ohio Republican who wants to kill the weapon in its gilded crib.

Ludicrous Strategy

One mission is to hunt Soviet mobile missiles—but right there, having written those words, "hunt Soviet mobile missiles," I must stop and take a little detour. With freedom breaking out in Eastern Europe, the abstract logic of nuclear strategy, always absurd, looks downright ludicrous. The idea of nuclear war between the superpowers has never seemed so divorced from history as it does today. Thus arguments such as I am about to advance—arguments against weapons systems which are based on strategic need and military efficiency—run the risk of looking beside the point. Since peace is at hand, why bother with these pre-peace categories? In an interview French President François Mitterand gave a good answer to that type of question when he said, "If Mr. Gorbachev were to fail, nothing can guarantee that a new Soviet power—which might not be Communist—wouldn't still be military and totalitarian." That new Soviet regime, it is necessary to remind ourselves, would still

Jack Beatty, "A Post-Cold War Budget," *The Atlantic Monthly,* February 1990. Reprinted with permission.

have more than 10,000 nuclear warheads targeted on the United States. Marshall Goldman, of Harvard's Russian Research Center, thinks that Gorbachev has at most only three or four years left; a high European official at NATO [North Atlantic Treaty Organization] headquarters predicts, "Unless he arrests the economic decline, he won't be there in two years." Cutting weapons systems on the grounds that Gorbachev has permanently halted the military competition between the superpowers might thus be letting the wish be father to the policy. We should cut them because we didn't need them before Gorbachev came to power, we don't need them with Gorbachev in power, and we won't need them if Gorbachev should fall from power.

"Trident II would force the Soviets to put their retaliatory forces on hair-trigger alert."

To resume: one mission for the Stealth is to hunt Soviet mobile missiles. Leaving aside whether that would be a mission impossible (it probably would), attacking the Soviet mobile missiles would be a mission undesirable. Each side needs an assured survivable force to secure deterrence. We have one in our ballistic-missile-carrying submarines; the Soviets have one in their land-based mobile missiles. To threaten the other side's survivable force is to raise the specter of a successful first strike in which you will disarm his only means of retaliation, putting him under remorseless pressure to use his retaliatory weapons against you first or lose them.

Once you see the fallacy of this so-called counterforce strategy—that putting your adversary's deterrent force at risk reduces your own security—the responsible course is not just to stop a particular weapons system but to abolish a whole function for future weapons systems. For, inevitably, Stealth will have its "follow-ons." Inevitably, leaks from the CIA [Central Intelligence Agency] will hint that the Soviets are fielding "a new generation of anti-Stealth weapons." Inevitably, there will be a "Stealth gap." And, inevitably, the taxpayers of the future will be called on to redeem the $80 odd billion already spent on Stealth with yet more billions for improvements, modifications, enhancements, or replacements for a system that should not exist to be improved, modified, enhanced, or replaced in the first place.

Insuring Americans

Or we could insure the medically uninsured. Some 30 million to 37 million Americans fall into that category, and approximately 15 million every year are denied medical care because they cannot pay for it. Most of them hold down the kind of low-wage, no-

benefits jobs that burgeoned in the 1980s. They are the people who deliver our papers, pump our gas, grill our hamburgers, carry our luggage, and care for our children. It is not their fault that they are trapped in sectors of the labor market that can't afford either to provide them with medical insurance as a fringe benefit or to pay them enough to insure themselves. And the work they do is socially necessary. Extending medical insurance to them as a form of social insurance would be a way of recognizing that. It would, to use a hoary word, be "just." It would also be expensive—estimates range from $25 billion to $50 billion a year. That sum would have to come out of taxes, but it would be only a small fraction of the $600 billion this society will spend on health care this year. And though it would be a dreaded "new social program," in the long run it would cost less than the system we have today. Society now does nothing for the pregnant teenage girl who avoids going to the doctor because she has no medical insurance. It patiently waits for her to present herself in the delivery room, and then spends $300,000 saving the life of her premature baby—a tragically shortsighted and profoundly wasteful result. Insuring the uninsured thus would strike a blow not only for social justice but also for economic efficiency.

The Trident II

We could continue to build and deploy the Trident II submarine-based ballistic missile. Tom Downey, a Democratic congressman from New York, says, "the Trident II will be the single most destabilizing first-strike weapon ever built." He may be right. Depending on how it is armed, each Trident II can be almost five times as destructive as the Trident I missiles now carried by our submarines. The Trident II will not only be superpowerful; it will be superaccurate, and thus able to hit and destroy Soviet missile installations. Its combination of destructive power and accuracy, when added to the quality of near-invulnerability conferred on submarine-based weapons, makes the Trident II a potential first-strike threat, one that would put Soviet nuclear forces under even greater "use it or lose it" pressure than Stealth. Trident II would force the Soviets to put their retaliatory forces on hair-trigger alert, and that would increase the danger of the only kind of nuclear war imaginable between the superpowers: an accidental one. From 1977 to 1984 there were more than 20,000 false indications of Soviet attacks on the United States; they must have had as many such indications from our side. Canceling Trident II would not only save $18 billion over ten years; it would absolve future taxpayers from the painfully unnecessary task of paying for Trident III.

Or we could return federal aid to education to its 1980 level in percentage terms. Back before the advent of the Reagan Administration the federal government devoted 2.5 percent of its total spending to education; in 1989 the amount was 2 percent, or $22.8 billion. President George Bush originally proposed increasing education spending by $441 million, which may sound like a lot of money but is in fact $110 million short of what Michael Milken made in salary in 1987. To return federal spending to the 1980 percentage, the President would have had to top Milken by $5.5 billion.

"We don't need Midgetman to deter a Soviet attack."

What could we accomplish in education by according it the same priority it enjoyed in 1980? We could, to begin with, fully fund Head Start, a program of enriched learning for poor pre-schoolers whose tonic effect on student achievement has been demonstrated in study after study for twenty years. Only 451,000 of the country's 1.7 million poor children are enrolled in Head Start. For about $1.2 billion more a year Head Start could be expanded to cover all eligible children for at least one year.

We could also serve every child eligible for aid under Chapter I of the Elementary and Secondary Education Act of 1965. Eight million children living in low-income census districts are theoretically eligible to receive the compensatory education called for under Chapter I, but in practice fewer than five million are getting it now. Under Chapter I, for $700 per child per year children who are at risk of repeating their grades receive remedial teaching. Every time a child repeats a grade, it costs the taxpayer $3,500, on average. Thus Chapter I doesn't just pay for itself; it saves the taxpayers money. Expanding Chapter I would cost nearly $3 billion. Former President Ronald Reagan has disparaged the idea that there is a convincing correlation between investment in education and a wider social gain. Yet a study done for the Committee for Economic Development found that money invested in education in fact paid off in the range of 7 to 11 percent after inflation.

Finally, if the level of federal aid to education were returned to what it was in 1980, more poor and middle-income young people could go to college. In 1979 Pell grants paid for 50 percent of a poor recipient's college costs, on average; now they cover only 29 percent. In a reversal of an encouraging trend of the 1970s, fewer and fewer young black men are going to college; the decline in federal support coupled with the rise in college costs is a big part of the reason why.

The Midgetman

We could build the mobile land-based missile known as Midgetman. In *The Atlantic Monthly* in 1989 R. James Woolsey, who is now President Bush's chief

conventional-arms negotiator, likened the Midgetman to a pair of suspenders backing up the belt of our submarines. In fact, we already have one pair of suspenders: our fleet of manned B-52 and B-1 bombers. How much should a man with a perfectly good belt (in congressional testimony a spokesman for the CIA said that that agency didn't believe the Soviets could deploy any effective threat to our submarines in the 1990s) and a perfectly good pair of suspenders be willing to pay for another pair of suspenders?

The projected cost of Midgetman is a sobering $30 billion plus. This small mobile missile is supposed to remove from the minds of Soviet planners any idea of mounting a first strike. The question is, Without the land-based mobile missile, are we vulnerable to such a first strike? Joshua Epstein, a defense analyst at the Brookings Institution, has calculated that even if the Soviets mounted a "perfect first strike"—one that destroyed *all* the 1,000 land-based missiles we have deployed, *all* the bombers on *all* our bases around the world, and *all* the missile-shooting submarines in port—the 50 percent of our submarines that are always at sea and the 30 percent of our bombers that are always on alert could still unleash more than 4,000 warheads on the Soviet Union. Having assumed the incredible in his worst-case thought experiment, Epstein goes on to posit the unimaginable. Suppose, he says, that the same Soviet air defense that could not stop a West German teenager from landing his Cessna in Red Square managed to mount a "perfect" air defense, knocking out all our bombers and all the cruise missiles they fired. In that worst of worst-case scenarios, 2,800 warheads from our missile-carrying submarines at sea would still fall on the Soviet Union. Epstein has asked senators and congressmen who favor Midgetman to tell him why the certainty of 2,800 warheads falling on the motherland is not enough to deter the Soviets. "They can't even name enough targets for the twenty-nine hundred warheads," he says, "yet they want to add more. We don't need Midgetman to deter a Soviet attack. We don't need it, period."

Aid to Poland

Or we could help to reinvigorate the Polish economy and give a fillip to Polish democracy. The Poles asked President Bush for $10 billion; he offered them $100 million. That pathetic response is a portent of America's decline as a great power. Poland, after all, is seeking to move from dictatorship to democracy, and from a command to a market economy. Two billion dollars from the United States now, coupled with the $8 billion in loans from the West as a whole that such a grant would make possible, would allow the Solidarity government to put the Polish economy on the path to self-sustaining economic growth. By giving the government the wherewithal to pay unemployment and resettlement allowances to workers displaced by the wrenching economic transition that Poland must undergo, an infusion of something more than what Senator Daniel Patrick Moynihan has derided as "walking-around money" would even help to legitimize democracy to the Poles. Yet a few million is all that an administration that wants to build the Midgetman and the Trident II and the Stealth bomber can do.

And we could cut cocaine production in Bolivia by 35 to 40 percent; that could be done with about $2 billion over three years, according to Jeffrey Sachs, a Harvard economist who has studied the problem. In 1986, before stopping drugs became the public's No. 1 demand of the federal government, Bolivia asked the Reagan Administration for money to finance a program of crop substitution and allied economic development. Pleading Gramm-Rudman limits on spending, Secretary of State George Shultz turned the country down. Suppose he had said yes. How much of the crack now tormenting mean streets from New York to Los Angeles would never have gotten into the country? Three hundred thousand Bolivian peasants work in the coca fields for their daily bread. The Bolivian government says that $2 billion from the United States, plus the loans thereby encouraged from others, would help it give those peasants a licit alternative to starvation. But the Bush Administration has asked Congress for only $261 million in military aid for the Andean countries (Peru, Colombia, and Bolivia) for fiscal 1990, and no economic aid at all until 1991.

And we could end the Third World debt crisis. That would take $5 billion, Jeffrey Sachs estimates. Under Sachs's plan the United States would withdraw $5 billion from the Treasury and place it in an account to guarantee interest payments to banks willing to make new loans to countries like Mexico and Brazil. But no money would actually be lost to the Treasury unless those countries defaulted on their loans, an eventuality that Sachs contends is unlikely. Thirty-nine countries with a total population of 850 million could thus be put on the road to recovery.

Strategic Defense Initiative

We could build the strategic defense initiative—at a great price not only in dollars (the research and development costs alone of SDI could run to $50 billion in the 1990s) but also in peace. The Soviets have agreed to go ahead with START [Strategic Arms Reduction Talks], an arms-reduction treaty that would cut the long-range nuclear-weapons arsenals of the superpowers by half. They are reserving the right to break out of that treaty, however, if the United States goes ahead with SDI testing or deployment that violates the antiballistic-missile treaty of 1972. This stipulation probably dooms the more robust versions of SDI that were floated a few years ago, because it is unlikely that a majority of congressmen would vote to fund a program that would cause the Soviets to break

out of a signed and ratified treaty. Even so, the House voted to spend $3.1 billion on SDI research in 1989, while the Senate voted $4.3 billion. In the promiscuous way of our politicians, they are apt to keep voting comparable sums for years, until SDI becomes a mortal threat to START, at which time they will pull the plug on SDI, having spent who knows how many billions to provide themselves with political cover and to ensure a steady flow of money from the political-action committees of defense contractors. Already SDI has absorbed $21 billion since Ronald Reagan launched the program, in 1983. That is enough. SDI was originally supposed to protect the U.S. population from Soviet attack. No one believes that is possible anymore. Now SDI is thought of as a system of partial defense for U.S. missile sites. It would strengthen deterrence, its proponents claim, by "complicating" Soviet attack plans. But do these plans need any more complicating? We have seen that even after a "perfect" Soviet first strike and a "perfect" Soviet defense against our retaliation, 2,800 warheads would be available to make the rubble bounce. Surely those 2,800 warheads constitute an inexorable complication. Congress should continue to fund research into the feasibility of defensive systems, but at about the rate devoted to this sort of deus ex machina before fear of Reagan and lust for PAC money made Congress back the escapist folly of SDI. A billion dollars a year should do it.

"SDI was originally supposed to protect the U.S. population from Soviet attack. No one believes that is possible anymore."

Would the elimination by this Congress of all funding for Stealth, Trident II, and Midgetman (along with its big brother, the rail-mobile MX: savings, $10.3 billion), and the reduction of SDI to the status of a modest research program, hamper the administration in the strategic-arms-reduction negotiations with the Soviets? Not to put too fine a point on it, no. The Soviets don't need the prod of U.S. bargaining chips to complete negotiations on START. Their collapsing economy gives them sufficient incentive to negotiate. In any case, it is most likely not the Soviets who have been holding up START but rather the U.S. Navy, which has had trouble making up its mind whether and how to ban nuclear-armed sea-launched cruise missiles. These so-called SLCMs, which, as noted in *The Atlantic Monthly* in February 1989, constitute one of the chief obstacles to START, are the bargaining chips of arms negotiations past—only they were not bargained away but retained, to bedevil future efforts to work out verifiable arms-reduction agreements. The Pentagon uses bargaining-chip arguments to co-opt the arms-control lobby into supporting new weapons systems. The congressman who votes for systems like Stealth and Midgetman on bargaining-chip grounds is using the cant of nuclear diplomacy to mask his real, pork-barrel motivations.

A Police Corps

With a billion of the $2 billion freed up by restraining SDI research, we could establish a national Police Corps. A Police Corps could add as many as 100,000 officers to overstretched police forces around the country. Enrollees would receive four years of guaranteed federal loans to cover college costs of up to $10,000 a year. In return, the 25,000 men and women selected each year, many of them members of minorities, would be expected to fulfill a four-year commitment to their local police force. When their term of service was over, the government would pay off their college loans. Since there are now 488,000 local policemen, the Police Corps would increase their ranks by an impressive 20 percent. "More significantly," Albert Hunt writes in *The Wall Street Journal*, "as the graduates would be placed almost exclusively on foot patrol, and not add to the police bureaucracies, the proposal should increase cops on the front lines by about 40%." The ratio of police officers to violent crimes has been tilting ominously toward violent crimes. In 1951, for example, there were 1,229 police officers and 361 violent crimes in the city of Buffalo, whereas in 1988 Buffalo had 970 police officers and 3,555 violent crimes. The strategic defense initiative will not defend the citizens of Buffalo. The Police Corps will.

We could continue to spend upwards of $150 billion every year defending Europe, the world's largest economic entity. Under the terms of the conventional-force-reductions talks, the Warsaw Pact would cut its forces in the Atlantic-to-Urals theater (Eastern Europe and the western military districts of the USSR) by 40 percent, but NATO would cut its forces by only 10 percent. That won't save the American taxpayer much money. Joshua Epstein, of Brookings, addresses that problem in a plan he has presented at the Pentagon and on Capitol Hill. He has discovered something that other commentators have missed about the talks: when forces designated for Europe but based in the United States are included, the proposed agreement will give NATO what NATO has never sought and does not need—*quantitative* superiority over the Soviets in two vital weapons categories, tanks and armored personnel carriers. NATO already spends $130 billion more than the Warsaw Pact countries every year to ensure that it retains its qualitative superiority in weapons. But NATO would need quantitative superiority only if it intended to invade Eastern Europe, and it has no idea of doing such a fantastic thing as that. Epstein recommends that NATO divest itself of this embarrassment of riches.

He would follow up the conventional-force-reductions talks with negotiations about a further 50 percent cut on both sides. He would then make a unilateral cut by demobilizing the National Guard reinforcement brigades for the forces we would be withdrawing from Europe. The 50 percent cut on top of the 10 percent cut plus the unilateral cut in the National Guard would save at least $20 billion, every year. And it would leave NATO "much better off" defensively than it is today, Epstein claims. That is because with each cut in the size of the Soviet forces, the chances of a successful blitzkrieg attack on Western Europe, a project fully as fantastical as a Western invasion of the East, rapidly approach the infinitesimal.

Rebuilding Our Roadways

With the $20 billion annually to be saved from NATO, we could make a down payment on the rebuilding of our infrastructure. It is in bad shape, our infrastructure. To see why, you have only to compare yesterday with today and the United States with Japan. In the 1960s we devoted 2.2 percent of all government spending to infrastructure, whereas by the late 1980s the level of investment had slipped to one percent. From 1973 to 1985 Japan spent five percent of its annual output on infrastructure, and it enjoyed an average productivity growth of 3.3 percent. Over the same years, the United States spent 0.3 percent on infrastructure, and it enjoyed—if that is the word—a 0.6 percent increase in productivity. Looking at these figures, David Alan Aschauer, of the Federal Reserve Bank of Chicago, concludes that government spending on infrastructure construction and maintenance spurs economic growth. And conversely: of the total U.S. productivity decline of 1.2 percent since 1970, fully one percent can be attributed to the neglect and deterioration of infrastructure, according to Aschauer. We need to invest $40 billion to $50 billion annually for a decade or more just to restore our extant physical capital. The Department of Transportation, to cite a resonant example, reports that the proportion of bridges that are defective rose from 10.6 percent in 1982 to 15.9 percent in 1989. Other estimates run as high as 40 percent. Maintenance work on bridges and on the interstate highway system will cost $300 billion over the decade.

"NATO would need quantitative superiority only if it intended to invade Eastern Europe, and it has no idea of doing such a fantastic thing."

Other vital infrastructure spending includes $25 billion for the modernization of the air-traffic-control system, $20 billion for the renovation of the existing production facilities, and $15 billion for state-of-the-art computers and a new long-distance phone system for the government.

The bridge that does not collapse, the air-traffic-control system that manages to land planes safely, the highway whose potholes do not break your rear axle—these prosaic achievements of government will not make any headlines. But in this essential though unheralded realm prudent investment could make an enormous difference to our children.

Bringing Soldiers Home

We could continue to station 31,000 army and 12,000 air force members in South Korea, at an annual cost of $2.6 billion, thirty-five years after the end of the Korean War.

Or we could withdraw 10,000 of them, saving $600 million every year. Senator Dale Bumpers, an Arkansas Democrat, has called for such a troop cut. Bumpers points out that South Korea now boasts a GNP seven times as big as North Korea's, and a population twice as big. With the trade surplus it runs with the United States every year, South Korea can afford to pay more of the cost of its own defense.

"All that stands between us and the more evenly prosperous future made possible by the ending of the Cold War is the depressing machinery of electioneering."

With the money we would save by bringing home some of our soldiers from South Korea, we could expand WIC, the supplemental-food program for women, infants, and children. Currently 7.3 million women are eligible for food and medical care under the income criteria of this program, but only 4.4 million are served by it. They are served well. For $40 a month a poor pregnant woman enrolled in WIC is given access to nutritious food as a medical prescription—in other words, to get the food, she has to see a health-care professional, who provides her with nutritional counseling and helps her find prenatal care. The program works: it increases the birth weight of babies. It also saves taxpayers money. Studies have shown that every dollar spent on the prenatal component of WIC actually saves three dollars in *that same year* since caring for a low-birth-weight baby in a neonatal clinic can run anywhere from $2,000 to $10,000 a day. WIC now costs $2.1 billion; for a billion more, it could be expanded to serve all the women eligible for it. . . .

We could invest in the programs outlined here, or we could put the billions saved from Stealth and Trident II and Midgetman and Star Wars and NATO and South Korea and COLAs [cost-of-living adjustment] and tax loopholes directly into deficit and

debt reduction. Each $50 billion reduction in the deficit would produce a one-point drop in real interest rates. It would also lower the trade deficit by $25 to $30 billion, because we would not need to borrow so much foreign capital to finance the budget deficit. Moreover, a $50 billion cut in the budget deficit would, by lowering interest rates, increase investment by $15 billion to $20 billion. That new investment would fuel economic growth, which in turn would lower the payout for unemployment compensation, welfare, and other expenses incurred by a sluggish economy. Higher growth, in its turn, would help us meet the interest payments on the national debt, which in 1989 drained away $240.86 billion, more than what the government spent for any program, including Social Security. There are only two ways of dealing with this senseless waste of money: put the budget into surplus and retire the debt, or make the economy more productive so as to increase the size of the revenue pie from which we now take 14 percent for debt service (it was 8.5 percent before the era of "fiscal conservatism" dawned in 1980). The second path is the only feasible one. And that is why decreasing the deficit to lower interest rates (and increasing investment in education, infrastructure, and research) is the key to a solvent posterity. All that stands between us and the more evenly prosperous future made possible by the ending of the Cold War is the depressing machinery of electioneering and simplification that has got our political system in its mindless, suffocating grip. The Republic can survive its problems. The question is, Can it survive its politics?

Jack Beatty is a senior editor of The Atlantic Monthly.

"The argument that the United States should put at risk . . . its national security . . . for some domestic programs now, fails to give sufficient weight to the consequences."

Defense Spending Should Not Be Cut to Fund Social Programs

Robert F. Ellsworth

The political mood in Washington is different from the popular "Don't Worry, Be Happy" song of 1988. The dilemmas that confront Washington will be hard to handle, in particular the defense budget dilemmas. How is the United States to:

—scale back the underfunded defense programs of the 1980s, to the tune of $230-475 billion, while continuing to maintain an adequate force structure and to modernize certain conventional and nuclear weapons systems, while limiting defense spending to help balance the budget deficit by mid-decade, without raising taxes or dipping into the Social Security trust fund?

—size and shape the American defense posture for an era when the traditional bases for planning may not be valid, or may not seem valid, given the "new thinking" in Moscow?

—deploy adequate forces where we will continue to have vital national interests—in Europe, Latin America, the Middle East and Persian Gulf, East Asia and the Pacific—without stretching ourselves so thin we lose credibility everywhere?

—reduce the role (and the numbers) of nuclear weapons without exposing the nation and its allies to the risks of overreliance on conventional forces for deterrence, which would be inherently incredible and destabilizing?

Political Game-Playing

These challenges are highly politicized by the subtle, high-stakes cross-games being pursued by the administration and congressional leaders: to co-opt each other to help fill in several large, fiscal sinkholes and end the eight-year Reagan freeze on innovative social benefit legislation without taking an initiative on taxes, yet maintain a strong defense posture along with existing international commitments.

Reprinted from *International Security*, vol. 13, no. 4, Spring 1989, Robert F. Ellsworth, "Maintaining U.S. Security in an Era of Fiscal Pressure," by permission of The MIT Press, Cambridge, Massachusetts.

There is an additional, deeper dilemma that the American defense budget process can only influence indirectly, but which will have enormous and pervasive influence on the defense budget, and it cannot be resolved before another five or ten years have elapsed. This is the strategic dilemma of how to guard against a new Soviet military threat come roaring back after a few years' breathing spell, while taking advantage in the meantime of Soviet President Mikhail Gorbachev's intention to replace East-West hostility and global class warfare with global trade, paranoia with openness, military superiority with "reasonable sufficiency," unremitting anti-Western hostility with cooperation on world hunger and the environment, and—above all—to replace the nuclear threat to the Soviet Union with the complete elimination of nuclear weapons by the year 2000. It is the cognitive dissonance between the five- or ten-year testing period for this new Soviet thinking, on the one hand, and immediate U.S. social and fiscal pressures coupled with the two-to-four-year American political calendar on the other, which makes this dilemma such a challenge.

We have seen several important changes in Soviet foreign policy and behavior as well as in domestic policy and practice. Gorbachev's December 7, 1988 U.N. [United Nations] speech promised even more. Indeed, the Soviets may have learned over the past fifteen years (as we probably have) that neither superpower can improve its national security by threatening the security of the other.

A "New" Soviet Union?

But so much remains the same in Soviet defense spending, arms production, and modernization. There have been some modest constraints on military exercise schedules, but no irreversible training changes. The Soviets maintain, under the redoubtable Marshal Ogarkov, now Commander of Forces in the Western TVD (Theater of Military Operations), a large, powerful, and modern conventional and nuclear-armed force in central Europe, capable of a short-warning assault. These forces may also provide blue-chip counters in any conventional arms reduction negotiations.

Since Gorbachev's Vladivostok speech on peace and friendship in mid-1986, the Soviets have not scaled back any of their major military deployments in Asia: the naval base at Cam Ranh Bay, ongoing militarization of the Japanese Northern Territories, an Air Army headquartered at Irkutsk including some 25 percent of the entire Soviet Backfire bomber fleet, and roughly half the total Soviet strategic ballistic missile-firing submarine fleet, which operates in the Sea of Okhotsk with supporting surface groups.

Beyond the two superpowers and their alliances, nearly every state in the world (as well as major criminal and terrorist groups) maintains armed forces to use as it sees fit. All have more and more access to nuclear and chemical weapons technology and to ballistic missiles and other systems for the delivery of warheads over long distances, as learned from the Iran-Iraq war. Local and regional wars will continue to occur. Some will involve the interests of the United States, some the interests of the Soviet Union, and some will involve the interests of both. Vietnam and Afghanistan may have slaked the superpowers' thirst for direct involvement with ground forces, but some local or regional clashes may require intervention by superpower air and naval forces.

"We should neither dismantle our force structure, nor hollow out our strength, nor cut back in Europe without major reductions and restructuring of Soviet forces."

In the light of actual conditions, therefore, it is difficult to sustain the argument that the United States has too large an overall military force structure. On the contrary, circumstances argue that U.S. military dispositions—in Western Europe and the Middle East, as in South and East Asia and the Pacific—reflect the minimum necessary (until further arms reductions are agreed) to assure deterrence and to deal with contingencies in the 1990s without the threat of using nuclear weapons in local or regional conflicts. No one wishes to see us with reduced global military potency, or with reduced negotiating strength in arms negotiations. Common sense resolution of defense budget dilemmas, therefore, will start with the proposition that we should neither dismantle our force structure, nor hollow out our strength, nor cut back in Europe without major reductions and restructuring of Soviet forces—but that nevertheless

we can economize to some degree on operations and maintenance for such forces. Second, this president and this Congress—either in concert or in opposition—must discipline our defense procurement programs to accord with the fiscal realities. Third, as we have on hand a 1988 budget authority of some $265 billion, we will be able to provide—with level budgets in real terms until 1994—necessary improvements of U.S. capabilities for the remainder of this century.

Above all, common sense tells us that there is no single way to save defense dollars: limitations on funding growth must be balanced across all budget categories—personnel, operation and maintenance, procurement—and among capabilities—strategic nuclear forces, land forces, tactical air forces, and ships. All U.S. forces and capabilities are meant to work with each other in a coherent whole, and that is the way the world perceives them. This perception should be reinforced, not weakened.

Budgeting for Strategic Nuclear Forces

At least for the remainder of this century, and most likely far beyond, the Soviets will continue to have the capability to wage nuclear war on the United States. Even under the most likely Strategic Arms Reduction Talks (START) agreement, both sides would be allowed to retain formidable strategic nuclear forces, to continue (if desired) all current nuclear force modernization plans, and to retain basic targeting doctrines unchanged. That is not to say the United States should actually procure and deploy everything allowed by START; on the contrary, media pressures to cut defense seem to focus on spectacular programs like the B-2 advanced technology bomber. Strategic force planning should take place under the aegis of an agreed concept of what the nation requires for long-term security. Such an agreed concept should provide for a modernized strategic command-and-control system, greatly reduced vulnerability of our offensive systems, and full coverage of necessary targets (which may be reduced in a START regime) by systems possessing high penetration capability. Uncertainty and incoherence have ruled too long.

In addition to the cost of the MX missiles and B-1B bombers that have recently become operational, further spending on strategic forces could mount to more than $236 billion during the decade of the 1990s if there were no agreed concept and if all the following were funded:

Advanced technology bomber	$64.0 billion
MX missiles	16.3
Rail garrison MX trains	9.1
Small ICBM	44.7
Trident II submarine	16.9
Trident II (D-5) missile	35.5
SDI (research and development only)	50.3
	$236.8

As well as deciding whether to go forward with some or all of these programs, however, the president and Congress need to decide what to do about our strategic command-and-control system. Despite recent improvements in survivability, this system—the president's only way of controlling, releasing, or limiting a U.S. nuclear response to a Soviet attack or to the threat of an imminent attack—may now be less than adequate to deal with the general problem of warning and response to a possible attack. Without more strategic warning time and a more reliable response capability, our National Command Authority could feel pressed to launch intercontinental ballistic missiles (ICBMs) on a hair-trigger, low-confidence warning. The Soviets, sensing this pressure, could in their turn feel pressed, in a period of mounting tension, to preempt early, and so on up the deadly, tightening spiral. A modernized system would allow much greater confidence, and safety, in deciding whether and when to launch. The cost savings (quite apart from the improvements in security and credibility flowing from reduced reliance on a hair-trigger reaction) could be very large. Both sides could cut back on the multiple echelons of reserve arsenals required to threaten retaliation against varieties of surprise attack.

Reducing Vulnerability

Greatly improved invulnerability of the U.S. land-based ICBM force (a serious silo-hardening program, or rail garrison deployment of MX, or small mobile ICBM, or some combination) would also help ease this tension, but the Reagan administration and four Congresses over eight years made very little progress along these lines. On the contrary, after spending $8-9 billion since 1980 trying to solve the silo-based ICBM vulnerability problem (the famous "window of vulnerability" of the late 1970s), U.S. ICBMs are still in silos and are still vulnerable. Common sense points in another direction: moving at once to new deployments.

Full-scale engineering development of a ballistic missile defense (BMD) could provide a hedge against the vulnerability of a mobile ICBM system, but an effective BMD system would be larger than that allowed by the 1972 Anti-Ballistic Missile (ABM) Treaty. Neither the fiscal costs of full-scale development of both the mobile ICBMs and BMD system, nor the political costs of Treaty renunciation, should stand in the way if this combination is clearly understood to be necessary.

With a modernized strategic command-and-control system, moreover, the president and the Congress might be able to increase our strategic warning time and improve our reliable response capability enough that the relative invulnerability of current (and START) land-based forces would be enhanced for many years to come.

In any case, the U.S. submarine-launched ballistic missile (SLBM) force probably should (and certainly could) be rapidly modified, mainly to provide greater dispersal of SLBM missiles among many more submarines than would be allowed under START counting rules, given the large number of missile tubes (24) per Ohio-class SSBN [ballistic missile submarine]. Initially this could be accomplished on a low-cost basis by effectively blocking several SLBM tubes on each boat. Later on, other more elegant solutions could be considered.

Budgeting for Conventional Forces

Nuclear weapons questions, important as they are, will not have as great an overall effect on U.S. defense budget dilemmas as will the much more expensive (by five- or six-fold) conventional force issues.

The chances of saving money through negotiated conventional arms reductions in the near term are slight. On the Soviet side, the political sensitivities —and even risks—of disrupting the current Five Year Plan before it is replaced with a new Plan (to run from 1991 through 1995) [and] the relative inflexibility of Soviet military doctrine. . . portend very slow going for any Soviet conventional arms reductions in the next few years. While we should not underestimate Gorbachev's ability to deliver big Soviet arms reductions if he can get a good deal from us, and while we should not hesitate to test the Soviets with proposals for substantial reductions, we for our part probably will not (and certainly should not) make unilateral cuts in our own force structure.

"The chances of saving money through negotiated conventional arms reductions in the near term are slight."

What does this leave for dealing with the budget problem? Congressional interests and some in the administration will prefer to save money by cutting into U.S. forces (the force structure does not enjoy strong political action committee [PAC] support). They will resist presidential moves to deal with the procurement underfunding problem by rescinding or stretching out major weapons procurement contracts, or by deferring planned new acquisitions (which do enjoy such support). Congress in any case does not have the authority to rescind contracts. Instead, the fastest and therefore most tempting way to reduce the fiscal deficit is by cutting heavily into defense budget authority for items such as pay, spare parts, and fuel, since these convert into outlays during the most immediate fiscal years. It is current-year outlays that affect the current-year deficit. The penalty for the nation, however, would be that its force structure would suffer, that readiness would deteriorate across the board, or both. As a result, U.S. international credibility would also deteriorate, probably to an even greater degree, as the perception would build on current expectations and would quickly spread, creating the myth of a "hollow" American military. A hollow American force posture would not be credible—as a deterrent to general war, as a set of bargaining counters in arms negotiations, as an intervention force in local wars, or for deployment on occasion as a demonstration of U.S. political commitment.

Better Ways to Save

A far better approach to saving substantial money in the defense budget while sustaining credibility would be to preserve total force structure and move several Army and Air Force units from active to reserve status. Then-Secretary of Defense Frank Carlucci already began in 1988 to reduce active duty military personnel and to increase the reserves. In terms of international perceptions, this would mean a visible decline in some units stationed forward (e.g., Korea) or that could be deployed quickly, and would therefore require high confidence that political leadership would respond to warnings of war in good time. It would require that all units, both active and reserve, be, and be seen to be, well equipped and trained.

This approach would also require some highly visible and well advertised beefing up of intercontinental mobility capability (airlift and fast sealift ships) and sustainability of U.S. forces in strategic locations (prepositioned matériel in Europe and the Pacific, facilities in the Middle East, and several more maritime prepositioning ships in the Indian Ocean). Intercontinental mobility capabilities are not great favorites of the high military cultures in this country, so getting the budget priorities these forces should have will not be easy. But it is one of the most important decisions to be made for the 1990s.

"A program to save substantial naval funds could include operation of aircraft carrier battle groups closer to the United States."

A program to save substantial naval funds could include operation of aircraft carrier battle groups closer to the United States. Adjustments could be made to keep several carriers close to home without undue risk, and at considerable economic saving. For example, carriers that take only five days to reach the Mediterranean from U.S. waters could prudently operate off the East Coast of North America, whereas those that take a month to reach the mouth of the Persian Gulf from California should retain home ports in Japan or the Philippines. This would also make possible the operation of some carriers with reserve air squadrons.

Other good ideas for saving money without materially degrading the international perception—and the reality—of U.S. military readiness and capability include using advanced electronic and software systems to upgrade the effectiveness of current platforms, instead of buying whole fleets of new platforms, using cruise missiles and remotely piloted vehicles to do larger shares of the tasks traditionally performed by manned aircraft, and using both simulators and specially instrumented training ranges to provide realistic training for groups and individuals at much lower cost and, in some cases, improved effectiveness and more quickly attainable readiness status.

Congressional agreement on procedures to let the Pentagon close some bases may save $2 to $5 billion by 1994. And it has been estimated that further implementation of the Packard Commission reforms could save $10 to $15 billion per year.

International Conditions

If the international environment does not improve on a long-term basis, and if the United States tries to economize now in ways that needlessly weaken force structure and alliances, we might have to spend even more money on defense in the late 1990s than we did in the early 1980s, and expose the nation to greater risks. But the United States is not a mere observer: we ourselves will bear a large share of the responsibility for the shape and the quality of the international environment. America does not lack for men and women of common sense, nerve, and will. The "decline" theorists argue that America could be brought down if our resources became overstretched. But even though the United States actually has adequate resources, we could also in some sense be brought down if we pursue ill-advised or short-sighted international policies, or if we have a minimalist attitude toward the world.

The argument that the United States should put at risk its international influence, its alliances, and its national security in the 21st century, for a marginal increase in resources for some domestic programs now, fails to give sufficient weight to the consequences of a deliberate pullback from its commitment to the international system. It also risks demeaning the quality and the purpose of American history and of America's destiny. If fiscal pressure leads to willy-nilly defense cuts now—and as a result we are confronted with catastrophe in the future—those howling for scapegoats in that future will be the very same "leaders" who now call for cuts and pullbacks in defense posture, no matter what.

Robert F. Ellsworth is vice chair of the International Institute for Strategic Studies, a London think tank. He formerly served as deputy secretary of defense and as U.S. ambassador to the North Atlantic Treaty Organization.

"If changes in the Soviet Union continue, under the best-case projections, the military can make far deeper cuts over the next decade without endangering Western security."

Improved Soviet-American Relations Warrant Defense Cuts

George J. Church

Generals and admirals for centuries have been notorious for planning to fight the last war. American military men are no different; for 45 years they have prepared for a Soviet version of the blitzkrieg. Panama, Grenada, Libya, even Korea and Viet Nam were all essentially sideshows. The Big One, if it ever came, would begin with the Warsaw Pact's tank and armored columns charging across the Fulda Gap into West Germany, starting a conflict that could escalate to a nuclear Armageddon. The effort to deter or defeat a Soviet invasion of Western Europe shaped almost everything about the U.S. military establishment: manpower requirements, weapons design, budget requests, the works.

With each passing day, this vision of the apocalypse becomes more archaic. The Kremlin's allies, if they can still be called that, are not only abandoning communism; they are demanding the removal of Soviet troops. . . .

The Soviets told the Poles that they are prepared to talk about troop reductions there. Torn by internal dissent and economic failure, the Soviet Union is in the process of unilaterally reducing its army by 500,000 soldiers.

The U.S., meanwhile, is left with a military strategy that was designed for a different world, and a force structure that must be not only reduced but also reshaped to avoid—or at worst, fight—the wars that America might actually get into in areas far from the Fulda Gap. How much and how fast are hotly contested subjects. Asked what he expected the U.S. military to look like in 20 years, Chairman of the Joint Chiefs of Staff Colin Powell referred to the dizzying pace of current events. "Twenty years?" he quipped. "I'm having trouble staying 20 days ahead right now."

George Bush acknowledged the rapid pace of events in 1990's State of the Union address as he called for the U.S. and the Soviet Union to cut their forces in Europe to 225,000 each, with no more than 195,000 of them in Central Europe. When the talks on conventional forces began in Vienna, 305,000 American troops still faced more than 600,000 Soviets. . . .

Already some American defense planners envision a further round of talks that would reduce U.S. and Soviet forces in Central Europe to as few as 100,000 a side. The defusing of this decades-old confrontation could result in the biggest demobilization of American forces, in Europe and elsewhere, since the end of World War II. The striking changes that began in 1989, Bush declared in his speech, "mark the beginning of a new era in the world's affairs."

Those hopeful words were reflected neither in the defense budget presented to Congress two days earlier nor in the somber assessments of some of the President's top advisers. Said a ranking defense official: "You could argue that a Soviet Union that has lost Eastern Europe, that feels it is under assault on the periphery, sees Azerbaijanis tear down the fence with Iran, has the Baltics trying to spin loose, faces unrest in the Ukraine, labor disturbances, and still possesses a marvelous military capability is a much more dangerous creature than we faced ten years ago under Brezhnev."

Such thinking seems curiously out of tune with the world as it looks in 1990. The Warsaw Pact, for all practical purposes, is dead as a military alliance. Soviet troops might have to fight their way through Warsaw, Prague and even Berlin before getting anywhere near the Fulda Gap, much less Bonn, Rotterdam or Paris. And while the Soviets were long considered capable of mobilizing for a strike at Western Europe in as little as 14 days, Pentagon analysts say that NATO [North Atlantic Treaty Organization] could now detect preparations a month in advance. Some outside experts argue that signs of war would be evident a full three months ahead of time.

Although Bush pointed out correctly that "we see little change in Soviet strategic modernization," even that dark prince of arms-control antagonists, former Assistant Secretary of Defense Richard Perle, has changed his thinking. "For the foreseeable future," says Perle, "I believe we can safely reduce the investment we make in protecting against a massive Soviet nuclear attack."

"The U.S. should push for a policy of minimal deterrence."

The "new era" the President spoke of will be dominated by economic competition more than military power. On that front, as Bush pointed out, the nation has a great deal to accomplish—restoring fiscal health, improving education standards, modernizing industry. Rethinking America's military needs is an important place to start. Former Defense Secretary Robert McNamara, now at the dovish end of the military spectrum, says the Pentagon's budget could be cut 50% by the end of the decade. "We could powerfully enhance our status as a world power, strengthen our military security, and redirect resources to more deserving sectors of our economy," he told *Time*.

The fundamental question for Americans is what military menaces they should be prepared for in the 1990s and beyond. And what kind of defense they will need to deal with such threats. A surprising consensus is emerging among planners in and out of Government.

Assuming further negotiated cuts in Europe, the U.S. will have either a far smaller force in Europe or none at all. Pentagon planners sensibly insist that initial U.S. troop and weapons cuts be reversible, so that American forces could return quickly in the unlikely event of a hostile Soviet move. "We need at least another year to determine whether the Soviet conventional restructuring is irreversible," argues James Blackwell, a military expert at the Center for Strategic and International Studies in Washington. This can be accomplished by having the Navy buy fast sea-lift ships that could transport U.S.-based soldiers to Europe in a crisis. The Air Force, similarly, should keep a powerful force of attack aircraft that could leap overseas on short notice. In addition, the military should maintain supply depots in Europe stocked with tanks, artillery and ammunition.

The superpowers should also re-examine their strategic nuclear forces, with the goal of achieving a far more stable balance. They should ban land-based multiple-warhead (MIRV) missiles, which are tempting targets for a first strike because an attacker can destroy the three to 14 warheads on such launchers by expending only one or two warheads of his own.

In addition, the U.S. should push for a policy of minimal deterrence. In the past ten years, the number of Soviet sites designated for nuclear destruction has grown to more than 20,000, including hundreds of bunkers and communications centers. The superpowers should evolve toward far smaller arsenals, designed merely to survive—and deter—a surprise attack with the capacity to retaliate.

In a world less dominated by superpower competition, however, both the U.S. and the Soviet Union may face unexpected challenges from increasingly well-armed Third World nations. The U.S. should be prepared for two types of action:

● Quick responses with limited force to sudden crises like terrorist hijackings.

● Somewhat more deliberate responses, but with greater force, to more complex situations like Panama.

The Navy should continue to play the central role in the global projection of U.S. might, though that should be possible with fewer aircraft carriers plus additional transport ships. It is also time for arms-control talks to be expanded to include reducing naval forces.

Given these opportunities, as well as the Pentagon's inescapable budget pressures, it is urgent that Washington devise a coherent plan to have an effective but smaller military by the end if not the middle of the decade. The Pentagon's typical gamesmanship—pretending to tighten its belt a little each year without rethinking basic issues—could lead to the worst outcome: a hodgepodge of cuts that will come anyway, guided not by foresight and leadership but by some of the worst instincts in politics.

"When your defense budget is not supported by a military strategy," says Congressman Les Aspin, chairman of the House Armed Services Committee, "it will be patched together with pork strategy." Each of the armed services, the defense industry and members of Congress will try to push major reductions off onto someone else while retaining as much as possible for themselves. Warns Phillip Karber of BDM Corp., a leading defense consulting firm: "If we do not set a direction of where our force structure can go, you can bet that we are going to end up paying more and getting less."

In that respect, Defense Secretary Dick Cheney's 1991 budget was all the more disappointing. Not only were his suggested cuts minimal, but the larger issues of military restructuring were tossed aside in the political jockeying over the proposal to close or scale back 72 military bases and installations. Cheney has appointed a task force to review the Pentagon's gold-plated strategic-weapons systems. But, notes Gordon Adams, respected director of the independent Defense Budget Project, "he did not even hint at slowing down any of them." These include the mobile MX/rail garrison missile project (budgeted for $2.8 billion), the B-2 Stealth bomber ($540 million apiece), and the

Seawolf submarine ($3.5 billion apiece), not to mention the Strategic Defense Initiative (which the Administration wants to increase from $3.6 billion to $4.5 billion next year).

Cheney's cuts in conventional weapons systems (such as the highly effective F-14D fighter plane) are mostly preludes to starting production on a new generation of weapons—such as the Advanced Tactical Fighter ($133 million apiece)—designed primarily for combat against the Warsaw Pact. Similarly, Cheney argues that all 14 of the Navy's deployable carrier battle groups would be useful in other global conflicts; never mind that the Navy initially lobbied for them by invoking the Soviet threat.

All told, Cheney's budget for Bush's "new era" would increase spending from $291 billion in fiscal 1990 to $295 billion in 1991; he argues that this amount, based on 4.6% inflation, is in fact a 2.6% decrease in purchasing power. Yet even adjusting for inflation, the 1991 figure would be nearly 30% higher than that in 1980, before Ronald Reagan began his anti-Soviet modernization buildup. The Bush Administration proposes to continue cutting at an inflation-discounted rate of 2% a year until 1995.

Respected military analysts, from the Brookings Institution's Lawrence Korb to Harvard's William Kaufmann, argue that if changes in the Soviet Union continue, under the best-case projections, the military can make far deeper cuts over the next decade without endangering Western security. The Pentagon, says Kaufmann, could save as much as 10% in 1991, 25% by 1995 and up to 50% by the year 2000. Some of these reductions—in Army divisions, in the Navy's outmoded battleships—would produce savings almost immediately. More significant cuts take longer because they involve the ships, planes and weapons scheduled to come on line over a period of years. Says Congressman Aspin: "Defense is by nature long-range planning. A decision you make today produces a ship in ten years." All the more reason to begin serious planning now.

There is no single way to cut the defense budget, but there are many obvious places for the Administration and Congress to start. If the following changes were made, the defense budget could be sliced by a third to a half over the next decade, falling as low as $150 billion (in current dollars) by the year 2000:

• The armed forces' 2 million manpower could be halved. The Army could shed three divisions immediately (rather than the two that Cheney proposed) and eight more of its present 18 divisions by 2000. The Army's troops alone could drop from 758,000 now to fewer than 400,000. Saving: $35 billion.

• The Army could reduce its inventory of tanks, artillery pieces and other weapons as part of the arms-control process in Europe. The 60-ton M-1 Abrams tank, in particular, was designed for massive armor

battles in Europe. It was of no use in the invasion of Panama because it is too big. Cheney has already recommended canceling future production. Saving: more than $6 billion by 1995 for the M-1 alone.

• The Marine Corps could be cut from three divisions to two in 2000, one based on each coast of the U.S. Saving: $1.2 billion annually.

• The bulk of U.S. forces could be stationed at home. Late in the decade U.S. forces will probably be completely out of South Korea and greatly reduced from the 50,000 currently in Japan. The U.S. has begun discussions with South Korea about withdrawing some 5,000 of the 43,000 American troops on duty there. Saving: $6 billion annually.

• The Navy could reduce its aircraft-carrier fleet from 14 to six—essentially one battle group apiece, plus replacements and training fleets, for the Atlantic, the Pacific and the Mediterranean. That would still allow it to fulfill its traditional assignments of keeping sea-lanes open, as in the Persian Gulf, or striking quickly at a distant foe, like Libya. But the admirals will have to give up former Navy Secretary John Lehman's "maritime strategy," which sought to send U.S. warships into Soviet waters to launch strikes against targets deep inside the U.S.S.R. Saving: $21 billion.

• With U.S. and Soviet nuclear warheads shrinking to half their present levels after a START [Strategic Arms Reduction Talks] treaty, the U.S. could press ahead for a ban on land-based MIRVed missiles. A ban would significantly favor the U.S. in numerical terms because the Soviets have far more of these monsters, such as the SS-18, which carries more than ten warheads. A MIRV ban would do away with existing U.S. missile systems like the ten-warhead MX and the triple-warhead Minuteman III. The cost of dismantling these existing systems would effectively cancel out the relatively small saving in operating costs. Saving: none.

"There is no single way to cut the defense budget, but there are many obvious places for the Administration and Congress to start."

• Trident submarines, with their new, highly accurate eight-warhead D-5 missiles, should be considered the firmest leg of the nuclear triad, offsetting any vulnerability of the land-based ICBMs [intercontinental ballistic missiles] and the huge cost of ever more sophisticated bombers. Even William Webster, the CIA's cautious director, has said that the Soviet Union will be "unable, at least in this decade, to threaten U.S. subs in the open ocean." But no new Tridents are necessary for the remainder of the '90s, and the U.S. should immediately kill the rest of the procurement program. Saving: $1.4 billion.

• Research for the Strategic Defense Initiative could be cut from $4.5 billion to $3 billion a year. This research should focus on developing technology, with no deployment necessary in this decade. Saving: $1.5 billion a year.

• Army attempts to build an antisatellite weapon would be put on hold. The U.S. depends far more heavily than the Soviet Union on satellites for intelligence and communications. It would have far more to lose in any competition with Moscow to see who could build the deadliest satellite killers. Saving: $1.4 billion.

Alas, all three services are still enamored of ultra-complex, ultra-expensive weapons systems. The argument used to be that only the highest of high-tech weapons could offset the Soviets' heavy superiority in numbers—no matter how suspect some of that Soviet power might have been. Now that the numerical superiority may be negotiated away, at least in Europe, the services are trying to find new arguments for the dollar devourers.

"Military strategists complain that they have to shape plans for a decade in a situation that changes explosively from week to week."

Making matters worse, the Defense Department is committed to spending $124 billion in the next decade for hardware such as the Navy's pricey ($60 billion for the program) A-12 attack aircraft, and the LHX helicopter, a beleaguered program that threatens to gobble up $42 billion over the next five years. Most of this money is not even anticipated in the current budget. As such programs are scratched or stretched out, the Pentagon faces enormous cancellation fees to contractors. Some of these weapons have already consumed millions of dollars in research and development.

One revolutionary approach to the usual research-develop-and-produce syndrome has been advocated by Aspin and backed by several experts outside the Pentagon. It is called "develop—but wait." Perform the R. and D. [research and development], in short, but go to production only if the imagined threat clearly emerges and if the cost is manageable. A more idealistic version advocated by Seth Bonder, president of a Michigan think tank called Vector Research, would encourage the Pentagon to invest in R. and D. but actually build new weapons only if they would correct an impending imbalance with the Soviet Union; it should pass up those that would give the U.S. a destabilizing military advantage.

This approach might justify not building more than one advanced Seawolf attack submarine. The reliable

Los Angeles class is still the best attack sub in the world, fully capable of protecting American vessels against enemy prowlers.

Similarly, the Navy and Air Force could slow the development of their new generation of advanced aircraft. The Navy's F-14 fighter, still in production, and the A-6 attack jet, which the Navy wants to phase out, are more than merely adequate. Nor does the venerable Air Force F-15 interceptor need to be replaced by a proposed Advanced Tactical Fighter. These grand projects could easily be kept on hold for ten years or more. The Air Force should also forget its new C-17 cargo plane, which costs $318 million, and stay with the long-proven and dependable C-141 and C-5.

Military strategists complain that they have to shape plans for a decade in a situation that changes explosively from week to week. But that danger is no excuse for not beginning to draw up a strategic plan to guide the reductions that a budget crunch is forcing on the U.S. no less than on the Soviet Union. Nor should it be allowed to obscure the happy prospects now beckoning Washington and Moscow alike.

An eloquent emphasis on the once-in-a-lifetime nature of the current circumstances was expressed by a career fighting man, General John Galvin, the American commander of NATO's unified forces. "If you're looking for the personification of the cold war, here I am," he said. "I'm seeing now the possibility that we can bring all of this to a close. If we can get 35 nations to sign on the dotted line on something that is irreversible and verifiable, and bring down the levels of armaments to a mere fraction of what they are today, then we really have achieved something that's worth all the sacrifices."

It is not often that a general shows such passion about cutting the forces under his command. That is but one indication of the historic opportunity facing America's political leadership. For once they should feel inspired to look ahead, not back at the last war.

George J. Church writes for Time, *a weekly newsmagazine.*

viewpoint 13

Improved Soviet-American Relations Do Not Warrant Defense Cuts

John Walcott

It was hard to watch the spectacle of the Red Army Chorus belting out "God Bless America" and "Turkey in the Straw" on the stage of Washington's Kennedy Center in December 1989 without realizing that the assumptions guiding America's military strategy for four decades are rapidly becoming as obsolete as muskets and men-of-war.

A volatile combination of political upheaval in Eastern Europe, escalating budget deficits in both the U.S. and the Soviet Union and potentially sweeping nuclear and conventional arms control agreements now promises dramatic changes in the two areas that have dominated American military thinking for 40 years: Deterring a Soviet nuclear attack and defending Western Europe against a massive buildup of Warsaw Pact men and armor. The tantalizing promise of an end to the cold war is already generating political pressure to bring American boys home from Europe and to slash the nation's $300 billion defense budget.

But even if it comes, the end of the superpower struggle will not mean the end of threats to America's global interests. . . . Guerrillas in El Salvador have launched an offensive to bring down that nation's U.S.-backed government. Small terrorist groups can still strike with ferocity, as West Germany's Red Army Faction reminded the world when it murdered the powerful chairman of the Deutsche Bank. Unstable and sometimes unfriendly nations are arming themselves with ballistic missiles and chemical weapons. Finally, as Lt. Col. Thomas Scott, a battalion commander in the Army's Seventh Light Infantry Division, observes, "Some of the greatest threats to American democracy, like drugs, are not things that can be identified as military targets. Multinational forces may have a greater role to play, as will peacekeeping forces."

Yet at a time when the swirl of events counsels prudence, political and budgetary forces are already pushing Congress and the Pentagon, willy-nilly, down a familiar path. Secretary of Defense Richard Cheney has ordered the services to plan for $150 billion in cuts from projected spending by 1994. The Army is considering eliminating at least 135,000 troops, reducing its active-duty divisions by one sixth; the Navy has proposed to abandon its long cherished goal of a 600-ship fleet, eliminating three aircraft carriers and about 60 other vessels. But Congress, which increasingly micromanages the Pentagon and its budget, is likely to consider that nothing more than a down payment. Representative Barney Frank (D-Mass.) puts it bluntly: "We're going to cut the hell out of it."

Redefining Strategy

From interviews with dozens of top Pentagon officials, officers and NCO s [noncommissioned officers] in the field, key members of Congress and defense experts and historians, there emerges a deep concern that if the nation is to avoid a dangerous military collapse, the inevitable budget cuts must be accompanied by a careful redefinition of America's defense strategy and priorities that re-examines fundamental questions in three areas:

• If strategic weapons are cut from the current 12,000 warheads on each side to much lower levels under new Strategic Arms Reduction Talks (START), how should much smaller U.S. nuclear forces be modernized and restructured to retain a credible deterrent?

• How large a U.S. conventional force should remain in Europe, and how should it be organized and equipped? To what extent can the U.S. and NATO [North Atlantic Treaty Organization] take advantage of new technologies and of earlier warnings of a possible Soviet attack?

• If containing Communism is no longer the overriding purpose of U.S. military power, how should the nation define which local or regional conflicts around the world endanger its national security and which do not? What sort of forces are needed to deal with the real threats in an era of limited resources?

Strategic Nuclear Weapons

There is one point on which military experts are in unanimous agreement: There is no way to uninvent nuclear weapons. Thus, the most important task of the U.S. military is still to prevent a nuclear holocaust. The cornerstone of U.S. nuclear strategy is maintaining an ability to survive a surprise attack on U.S. bomber bases and missile silos with enough nuclear weapons to launch a counterattack on some 2000 militarily significant Soviet targets.

Reducing strategic weapons to levels substantially below those contemplated in a START agreement—which would leave about 9,000 warheads on each side—poses enormous complications. According to George Bing of Lawrence Livermore National Laboratory in Livermore, Calif., calculations show that if both sides cut their nuclear forces to 6,000 warheads, "we could still accomplish the military missions we believe we can today"—that is, hitting the 2,000 Soviet military targets.

To cut forces to 3,000 warheads each and still maintain a credible deterrent, however, would require fundamentally restructuring U.S. forces so they are better able to ride out a Soviet attack. "It could be mobile missiles, submarines or basing aircraft so they're more survivable," says Bing.

"The nuclear forces of France, Britain or even Israel could start to become significant factors in the global balance of power."

A report by William Kaufmann of the Brookings Institution, which maps out a 50 percent cut in the defense budget, suggests that a minimum survivable force of 4,000 warheads that meets these constraints would save $30 billion a year, although it would rely heavily on submarines, which cannot respond as quickly as land-based missiles. But some analysts worry that current START proposals, coupled with the fact that the U.S. decided some time ago to build a smaller number of larger nuclear submarines, could cut the ballistic-missile-submarine fleet to 18 or fewer Trident subs. Such a small U.S. missile-submarine fleet could become vulnerable to Soviet attack submarines and improving anti-submarine-warfare technology, neither of which would be limited by any arms treaty now in prospect.

A study by the Center for Strategic and International Studies suggests that substantial savings would still be possible if the $30 billion, single-warhead Midgetman mobile missile were added to create a "surprise-resilient force" that would be less vulnerable to a Soviet attack than today's silo-based missiles. The Midgetman force, which would be carried on trucks, could be quickly dispersed upon warning of a Soviet attack.

But at very low levels of strategic weapons, the nuclear forces of France, Britain or even Israel could start to become significant factors in the global balance of power; the danger that one side could obtain a decisive advantage by cheating is also much greater when each side has only a few hundred weapons. "We're going to live in a world of deterrence as far as we can see," says former arms negotiator John Rhinelander. "We've lost enough plutonium in the pipeline that we can never be sure there isn't one [bomb] out there."

Conventional Arms In Europe

Changes in U.S. strategic nuclear forces will be shaped by a complicated START agreement that will limit both sides' forces. But in the heart of Europe, a political revolution threatens to outrun negotiators from 23 nations who are hammering out a Conventional Forces in Europe (CFE) agreement slashing NATO and Warsaw Pact forces.

An official re-evaluation of the Soviet military threat underscores the magnitude of the change. The Soviets already are withdrawing 240,000 of their 600,000 troops from Eastern Europe, and at least 10,000 more could be pulled out under a CFE agreement. But a classified U.S. Intelligence study revealed by *The Washington Post* all but rules out the possibility of a Soviet surprise attack in Europe. The study concluded that the U.S. would have at least a month's, and more likely two to six months', warning of a Soviet attack, given the low state of readiness of most Soviet forces.

If non-Communist or coalition governments can take and hold power in Hungary, Poland, Czechoslovakia and East Germany, and if the U.S.S.R. withdraws all or even most of its troops from those countries, the former Soviet bloc will be transformed from a staging area for the armored blitzkrieg the West has long dreaded into a *cordon sanitaire* 500 miles wide, benefiting NATO and the Soviet Union alike. Already, says former Defense Secretary James Schlesinger, "one can scarcely look at Polish, Hungarian, Czechoslovak or East German divisions as even being available to the Warsaw Pact."

Big Soviet troop withdrawals, however, almost certainly would stimulate political pressure in both Europe and the U.S. for American troops to retreat across the Atlantic while the Soviets march home across the Polish border. . . .

Britain, France and West Germany, however, all oppose more rapid cuts in manpower, which could hobble NATO's longtime strategy of "foward defense":

There simply would be too few soldiers to fill all the gaps and guard the East-West frontier, even against a smaller and less heavily armored attack. That, in turn, could weaken the credibility of the NATO Alliance, especially in front-line West Germany. But some defense experts argue that a host of new antitank and other technologies—in which the West has a comparative advantage—would enable a relatively small NATO force to defend the border, especially if a CFE agreement forces the Soviets to destroy 25,000-30,000 tanks.

In the view of a growing number of military experts, the increased warning time of Soviet mobilization now allows the U.S. Army to shift a substantial amount of heavy armor out of highly ready forces into the reserves. The Army maintains four full-strength armored and mechanized divisions in West Germany and 14 more active-duty divisions in the U.S., six of which are ready to move to Europe within 10 days of a mobilization order. Altogether, some 300,000 U.S. troops, out of a total force of 2.1 million, are stationed in Europe. Those forces, plus the six rapid-reinforcement divisions, account for about one third of the $300 billion defense budget. Former Secretary of the Navy John Lehman, in an article for the Reserve Officers Association, suggested transferring one third of all active-duty forces to the reserves, at a savings of as much as $45 billion a year. In place of 18 active and 10 reserve Army divisions, Lehman suggests just the reverse.

The New Threats

The newest, and mostly unmet, challenge for U.S. planners is to devise ways to counter the multitude of threats, from terrorist attacks to the disruption of Persian Gulf oil supplies, that are emerging to replace the single "evil empire." The first order of business in this age of military limits is deciding which threats demand U.S. attention and which can safely be ignored. The oil-rich Persian Gulf, nearby Central America, and the Philippines, where the huge Clark Air Base and Subic Bay Naval Base are located, all seem more crucial to U.S. security than, say, civil wars in Cambodia or in the Horn of Africa.

Although regional conflicts may seem less significant once the rationale of stopping Communist expansion is gone, "potential adversaries in the Third World are no longer trivial military problems," warns Under Secretary of Defense for Policy Paul Wolfowitz. "With the Soviet Union, the U.S. could spend years looking at scenarios," says Seth Carus, a research fellow at the Naval War College Foundation. "In the Third World, every threat will be unique."

The armies of at least 12 Third World countries are equipped with more than 1,000 tanks. During the Iran-Iraq War the combatants fired more than 1,000 missiles at each other's cities. The Afghan Army fired some 1,000 Scud surface-to-surface missiles against *mujeheddin* guerrillas just in one year.

Iran and Iraq—both sworn enemies of Israel, America's closest Mideast ally—are continuing to develop ballistic missiles today; according to the CIA [Central Intelligence Agency], at least 15 developing countries could be producing missiles with ranges of 3,000 miles by the turn of the century. Several of the countries on the list, notably Syria, Iran and Iraq, overlap with the list of 20 or so countries that are also believed to have chemical weapons either in hand or under development.

"At least 15 developing countries could be producing missiles with ranges of 3,000 miles by the turn of the century."

Faced with such a diffusion of possible threats, Pentagon planners are stressing the need to build a much more flexible U.S. force structure that can operate anywhere in the world. . . .

These flexible forces will need good information, and keeping track of new areas of concern in the Third World may prove especially difficult. Even the latest all-weather, billion-dollar U.S. spy satellites are not capable of collecting much intelligence on tiny terrorist cells or coup plots that are invisible and inaudible from space; in any event, American reconnaissance satellites are so tied down keeping tabs on Soviet missiles that they are unable to provide much coverage elsewhere. That was one reason the U.S. failed for more than a year to detect Chinese-made surface-to-surface missiles that Saudi Arabia deployed in 1987.

But U.S. intelligence agencies still are not well equipped to put agents on the ground in the Third World. Intelligence failures underscore this shortcoming. During the U.S. invasion of Grenada, the landing forces had no proper maps and landed where a Cuban force was encamped. In the Persian Gulf, it was only after the *Stark* was hit by *two* Exocet missiles that the intelligence agencies learned the Iraqis had modified their French Mirage F-1s to carry more than one missile.

War Isn't Over

The most serious wounds to the U.S. military, however, could be self-inflicted. In Congress, the parochial self-interest of competing states and districts often determines how and where budget cuts will be made. While Defense Secretary Cheney and JCS [Joint Chiefs of Staff] Chairman Colin Powell have said they intend to restructure the military rather than just make across-the-board cuts, Cheney so far seems to have taken the path of least resistance, leaving it up to the services to squeeze and trim within existing plans. "As far as I can see, it's one third, one third, one third," says Adm. William Crowe (Ret.), former chairman of the JCS, referring to how the cuts will be apportioned among the Army, Air Force and Navy. "There ain't gonna be any strategic rationale."

Such cutbacks in the 1970s produced so-called hollow forces as munitions, maintenance and training were squeezed. "If you don't have a strategy, you try to do everything, just do it slower," says former Secretary of Defense Harold Brown. "And if you try to make deep cuts that way, you can wind up with nothing but still spend almost as much money." The B-2 Stealth bomber is a telling example: The House Armed Services Committee has estimated that a proposal to make the program more "affordable" by cutting annual purchases in half would, in the long run, double the cost of each plane—to $1 billion.

"The most serious wounds to the U.S. military . . . could be self-inflicted."

The lesson is that although the central preoccupation of U.S. military strategy for the last 40 years may be waning, war isn't over. U.S. commitments to collective security and deterrence, argues military historian Allan Millett of Ohio State University, make the present period unlike any other in the nation's history. "We now have interests in the world that are going to require a higher level of readiness than we have *ever* had," he says. But amid all the excitement and the pressure for change, that will require America to remain as cool under pressure as it has always expected its soldiers to be.

John Walcott is an assistant managing editor for U.S. News & World Report, *a weekly newsmagazine.*

The Cold War's End Proves the Benefits of Arms Control

Joseph S. Nye Jr.

For the past 30 years, arms control has been central to the U.S.-Soviet relationship. Now, if the cold war is over, what will be the role of arms control? On the one hand, the relaxed political climate improves the prospects for reaching and ratifying agreements. On the other, improved U.S.-Soviet relations also reduce anxiety about nuclear weapons and urgency about arms control initiatives. Polls show that Americans are now more worried about the state of the U.S. economy and drugs than about the Soviet Union and threat of nuclear weapons.

Geopolitical analysts warn about the diffusion of power in world politics; the spread of chemical and ballistic missile technologies to some 20 nations in the next decade will pose a new type of security threat. Some critics assail the Bush Administration for moving too slowly on the traditional bipolar strategic arms control agenda; others call for giving a higher priority to proliferation and multilateral measures. Still others continue to warn that all arms control agreements are a snare and a delusion. The new twist for the time ahead is that the United States and the Soviet Union, though remaining antagonists on the traditional agenda, will find themselves to be partners in some of the emerging problems of arms control.

The Merits of Arms Control

The 1980s began with a sharp debate about the merits of arms control. Many officials in the Reagan Administration contended that arms control was more of a problem than a solution; it lulled public opinion in Western democracies into accepting Soviet strategic superiority. While the pressure of public opinion brought the administration back to arms control negotiations within its first year, little was accomplished until 1986. Then, in its last two years, the Reagan Administration signed an Intermediate-

Joseph S. Nye Jr., "Arms Control After the Cold War," *Foreign Affairs,* vol. 68, no. 5, Winter 1989/90. Reprinted by permission of FOREIGN AFFAIRS, Winter 1989/90 by the Council on Foreign Relations, Inc.

range Nuclear Forces (INF) agreement, causing consternation among many of the president's most ardent supporters. In addition, the administration made substantial progress toward a treaty in the Strategic Arms Reduction Talks (START).

Some conservatives were outraged by Ronald Reagan's new views: Howard Phillips called him a speech-reader for appeasers. But other conservatives base their skepticism on a more general critique of arms control. Irving Kristol, for example, has reiterated his charge that arms control agreements do not lead to enduring settlements of conflicts but instead

> tend to be slow, tedious and conducted in an atmosphere of skepticism and suspicion. As a result, agreements are likely to have limited scope. Moreover, technological innovations in weaponry, to say nothing of changes in national leadership, will always make an arms control treaty vulnerable to conflicting interpretations or outright indifference.

Some aspects of this case against arms control have merit. Weapons are symptoms rather than basic causes of hostility. The legalistic approach to seeking compliance with treaties can lead to disproportionate responses in a period of extreme distrust. Arms control negotiations may sometimes slow the process of change. For example, since 1980, NATO [North Atlantic Treaty Organization] has reduced short-range nuclear weapons based in the front lines of Europe. It is quite plausible that efforts to negotiate these reductions in the context of formal arms control agreements would have hindered this stabilizing change. Similarly, some Soviets say that Mikhail Gorbachev announced a unilateral reduction of conventional forces in 1988 at the United Nations rather than at the bargaining table for fear that negotiations would slow the process.

Even some of the founding fathers of modern arms control, such as Thomas Schelling, have expressed skepticism about too much reliance on formal agreements. Many strategists believe that types of weapons, their vulnerability and their susceptibility to central control are more important than their numbers. Reductions are not good per se, but must be judged in light of these characteristics. At low numbers, deceptive practices, hidden weapons and breakouts from treaty constraints could have a greater impact on security than at higher levels. However, given the high levels of existing arsenals, reductions would have to be much deeper than currently foreseen before such factors become a serious security problem.

Critics' Fears

A careful study of the U.S.-Soviet arms control record in the pre-Gorbachev era concluded that critics' fears that arms control agreements would lull the public and weaken the defenses of democracies have not been borne out. On the other hand, the hopes of proponents that arms control would save money and lead to dramatic reductions were not borne out either [as Albert Carnesale and Richard Haass argue].

> What emerges above all is the modesty of what arms control has wrought. Expectations, for better or worse, for the most part have not been realized. The stridency of the debate, however, provides little clue to this modest reality. . . . If the history reveals anything, it is that arms control has proven neither as promising as some had hoped nor as dangerous as others had feared.

In the past three decades, arms control agreements were concluded only when neither side had an appreciable advantage; agreements were not reached when either side had a strong preference for development of a new weapon. Based on this modest record, critics argue that arms control contributes little to international stability.

Critics, however, miss the point: arms control is part of a political process. Too often the experts judge arms control proposals on their technical details rather than on their political significance. For example, the INF agreement was militarily insignificant. In fact one could argue that, in terms of stability, by first removing longer-range nuclear missiles from Europe rather than starting with the short-range artillery, the INF agreement seized the wrong end of the stick concerning command and control of weapons during crises. But the political significance of the INF agreement—the improvement in the U.S.-Soviet relationship in the second half of the Reagan Administration—far outweighed the technical problems related to the details of military doctrine.

A Reassuring Process

Arms control reassures the publics in Western democracies. The process is an inevitable and important part of the political bargaining over defense budgets and modernization. Whatever the strategists

may say, the public cares about reductions in numbers because it is difficult to grasp other measurements or present them in clear political terms. Numbers matter because they are a readily perceived index. Since a major benefit of arms control is domestic political reassurance, general ceilings or reductions can make important contributions to security and force planning even if the number of weapons alone is a poor measure of the risk of nuclear war.

Arms control also provides reassurances to adversaries. In a sense, all of arms control is a confidence- and security-building measure. By increasing transparency and communication among adversaries, worst case analyses are limited and security dilemmas are alleviated. It may be that the most important aspects of the two Strategic Arms Limitation Talks (SALT) agreements were the provisions on open skies for satellite reconnaissance, the agreed counting rules for various types of weapons and the establishment of a Standing Consultative Commission to discuss alleged violations and misunderstandings. In that sense, informal operational arms control and formally negotiated reductions are not exclusive alternatives; they can complement one another.

"Arms control reassures the publics in Western democracies."

The agreements on incidents at sea, crisis centers, confidence- and security-building measures in Europe and the agreement on the Prevention of Dangerous Military Activities have been scorned by some experts as the "junk food" of arms control. But the classical distinctions between reductions in arms and measures to build confidence and security have begun to blur. Both structural and operational arms control are parts of a larger process of political reassurance among adversaries.

Arms Control's Political Role

Skeptics might reply that despite the past political role of arms control, the current climate of U.S.-Soviet relations makes further arms control unnecessary because the public and the Soviets no longer need such reassurance in the Gorbachev era. But such a reply fails to understand the institutional role of arms control agreements. As arms control agreements become accepted, political leaders and bureaucratic planners on both sides are less likely to base their strategy upon far-fetched worst case scenarios. Arms control and defense plans tend to reinforce each other. Despite its rhetoric in the early 1980s, the Reagan Administration was better off staying within the framework of the two SALT agreements because they constrained the Soviets and there was little that the

United States could build in the short run. To take a different example, in November 1983 the Soviets walked out of INF negotiations with the United States to protest new NATO deployments, yet they still continued to meet with the United States to discuss nonproliferation via institutions that had been established when relations were easier.

In the 1950s the early theorists of modern arms control aimed to reduce the risk and damage of war and save resources. Since those goals are not very different from the objectives of defense policy, it is natural for defense and arms control measures to interact as complementary means to the same ends. As a pattern of reciprocity develops, both sides begin to redefine their interests. Even in the pre-Gorbachev era, as dour a figure as Andrei Gromyko reportedly lobbied to include rising young Soviet officers in SALT delegations on the grounds that "the more contact they have with the Americans, the easier it will be to turn our soldiers into something more than just martinets."

The opportunities presented by the current political climate and the possibility of a return to cold war relations reinforce the argument for reaching good agreements now: their institutional effects will linger and ameliorate our security problems if relations deteriorate in the future between the United States and the Soviet Union. While Gorbachev's glasnost may have increased transparency and communications beyond the arms control process, both formal reductions (such as the asymmetrical reductions in conventional forces in Europe) and informal agreements that provide access to information (such as exchanges among military officers and visits to Soviet facilities) can help lock in gains for Western security.

International Security Regimes

Skeptics also neglect a further political role of arms control: the establishment of international security regimes. By treating the military relations among states as a problem of common security, arms control agreements help legitimize some activities and discourage others. These international regimes cannot be kept in separate watertight compartments. For example, the long-term management of nuclear proliferation would be impossible in the context of a totally unconstrained U.S.-Soviet nuclear arms race. Similarly, it is difficult to imagine the United States and Soviet Union managing the diffusion of chemical and biological weapons technology if these two countries were engaged in unconstrained developments in those fields.

Skeptics point out that most states develop nuclear and chemical weapons because of security problems with their neighbors, not because the United States or the Soviet Union promise to disarm. This argument is largely correct. The existence of Article Six of the Nonproliferation Treaty (NPT), in which the

superpowers promise to reduce their arsenals, did not deter Pakistan, South Africa or Israel in their nuclear policies. On the other hand, a renewal of the NPT in 1995 will be harder if it must take place in the context of a sharp U.S.-Soviet nuclear arms competition. As one looks further down the road and contemplates the diffusion of destructive power and technology to poor countries and transnational groups, bilateral U.S.-Soviet arms control cannot be divorced from the multilateral arms control problem. On the contrary, the management of international security in the future is likely to require more, not less, attention to the political role of arms control.

"Both formal reductions . . . and informal agreements that provide access to information . . . can help lock in gains for Western security."

If measured in terms of numbers of strategic warheads, Soviet missile accuracy and the vulnerability of U.S. intercontinental ballistic missiles (ICBMs), the strategic balance actually worsened during the Reagan years. So why the striking changes in the American arms control policy in the 1980s? Reagan's military budgets, his rhetoric about the Strategic Defense Initiative and the INF deployment in Europe all played a role, but the primary answer is Mikhail Gorbachev. Immediately after coming to power in 1985, Gorbachev altered the Soviet stance in strategic talks by conceding the validity of U.S. concerns about the vulnerability of fixed land-based missiles. In 1986, he permitted intrusive on-site inspections related to confidence- and security-building measures in Europe. In 1987 Gorbachev agreed to an INF agreement largely on Western terms, and in December 1988 he announced unilateral cuts in Soviet conventional forces. In 1989 he proposed deep asymmetrical cuts in conventional forces in Europe [CFE]. . . .

The Post Cold War Era

In the post cold war era, arms control may lead to major reductions in the forces of the superpowers. Even in this new era, however, military forces will still be needed because of the normal course of great power politics and because of the new diffusion of destructive power. Moreover, there is always the prospect that the changes in the Soviet Union could be reversed.

The United States can protect itself against reversal in three ways. First, it should seek those reductions that not only enhance U.S. security, but that take time for a new Soviet leadership to restore. Second, the United States should seek verification and inspection procedures that, if violated by a new Soviet leader, set off clear alarms. Third, it should seek procedures for informal visits and consultations that reinforce groups dedicated to glasnost in the Soviet political and military leadership. Thus, careful attention to detail is needed to make sure that dismantling the military edifices of the cold war does not create technical instabilities that would reduce or threaten U.S. security in the future. At the same time, technical details should not blind Americans to the larger political roles of arms control.

It has become fashionable to speak of the end of the cold war and even of the "end of history." When the cold war ended may well be a matter of semantics: strictly defined, as a period of intense hostility and little communication, it probably ended in the 1960s. The "little cold war" of the early 1980s was mostly rhetorical. More important is the fact that the cold war and the division of Europe produced four decades of relative stability, albeit at a high price for East Europeans.

Rather than the end of history, we are now seeing the return of history in Europe with its ethnic tensions and the unsolved problem of Germany's role, which Bismarck put on the international agenda in 1870. The peaceful evolution of new arrangements in Europe will make reassurance more necessary than ever. Arms control can play a large part in that reassurance by reducing and restructuring force levels in the conventional arms negotiations and by establishing a variety of confidence-building measures in the Conference on Security and Cooperation in Europe. Negotiating gradual changes in the overall European security framework can help to alleviate the anxieties and overreactions that would otherwise derail economic evolution and integration of the continent.

The United States can use this period to accumulate security gains, banking them against a possible future downturn or reversal in U.S.-Soviet relations. Because the institutional effects of arms control tend to continue, an important goal now for the United States is to lock in the benefits of the Gorbachev era: force reductions in START and CFE, as well as verification and consultation procedures that build transparency and regularized communication. In turn, this broad process may help to reinforce the changing security concepts in the Soviet Union. Agreements create domestic effects there as well as in this country.

Improved Bilateral Relations

The current period of improved bilateral relations provides an important opportunity for the United States and the Soviet Union to work together with other countries to reinforce and establish regimes for dealing with the diffusion of power. Here too there are gains to be locked in. Such multilateral arms control regimes will have to be considered in a broader context of security. For example, the superpowers have already rediscovered the value of

U.N. [United Nations] peacekeeping forces, itself a confidence- and security-building measure. They may also rediscover the wisdom of the early postwar architects of the U.N. Charter and particularly of the U.N. Security Council. In an era when the great powers are reducing their involvement in the Third World, other countries may develop greater interest in measures for the regional constraint of force. There are signs that some of the less developed countries have begun to understand that their traditional litany of complaints are somewhat beside the point. As one U.N. diplomat put it to me privately, "What are we going to do when we don't have the great powers to kick around any more?"

"The management of international security in the future is likely to require more, not less, attention to the political role of arms control."

There are several implications for U.S. policy about the new role of arms control in a post cold war period. The United States will have to pay more attention to the multilateral dimensions of arms control, and more attention to the relationship between bilateral and multilateral arms control. It will have to pay more attention to the relationship of arms control to regional political processes. And the United States will have to pay more attention to how arms control relates to other instruments and other goals in U.S. foreign policy.

Arms control will never provide all the answers to national security. In some cases, it might even do more harm than good. In all cases, it will have to be integrated with other dimensions of policy and other policy instruments. But the changing nature of world politics suggests both new roles and new importance for arms control. If an arms control process did not exist, we would assuredly have to invent it.

Joseph S. Nye Jr. is Ford Foundation professor of international security at Harvard University in Cambridge, Massachusetts, and the author of Bound to Lead: The Changing Nature of American Power.

"Future historians will come to view formal arms control as just another accoutrement of the Cold War."

The Cold War's End Proves the Irrelevance of Arms Control

Kenneth L. Adelman

In 1932, the Spanish Ambassador to the Geneva Disarmament Conference told this fable in response to the new Soviet foreign minister, Maxim Litvinoff, grandly proposing "total and general disarmament":

> *"When the animals had gathered, the lion looked at the eagle and said gravely, 'We must abolish talons.' The elephant looked at the tiger and said, 'We must abolish claws and jaws.'*
>
> *"Thus each animal in turn proposed the abolition of the weapons he did not have until at last the bear rose up and said in tones of sweet reasonableness: 'Comrades, let us abolish everything—everything but the great universal embrace.'"*

Why this topic? Because concerns about relations between the superpowers and about nuclear weapons have filled our thoughts and aroused our passions for the past forty years. The specter of nuclear war is hanging over our civilization, and we search for ways to make it less haunting, if not to make it vanish altogether.

Why now? Because we've come to romanticize United States-Soviet summits and idolize arms accords in a way that is both dangerous and inaccurate, and because we hear the sound of Cold War ice cracking. When George Bush sits down across the table from Mikhail Gorbachev—surely the most imposing, impressive person on the world stage today—he will face the same man Ronald Reagan [met]. For the most part, they will discuss the same issues.

Yet everything outside that room will be different. Indeed the watchword of our times, at least in this realm, may have first been uttered by Dorothy: "Toto, I have the feeling we're not in Kansas any more." A sweeping transformation in big-power politics began during the Reagan administration, a change which

may eventually be considered Reagan's main historical legacy.

Why the skepticism? Our hopes for productive arms accords have soared with the presence of a "new thinking" Soviet leader (after a string of "no thinking" ones); with the signing of an INF [Intermediate-Range Nuclear Forces] Treaty; with the growth of greater trust and stronger economic incentives. But prospects for a meaningful arms accord are actually worse than ever. Traditional arms control has reached a dead end.

This is not particularly sad. Greater security and fewer nuclear weapons—the goals arms control has generally failed to achieve—will, alas, be achieved.

The same factors which *hurt* chances for traditional arms control actually *help* chances for a safer world. Benefits commonly ascribed to—but seldom delivered by—traditional arms control can now be advanced in other ways and with more success.

To wit: for thirteen years in Vienna, armies of diplomats gathered to negotiate conventional arms reductions before utterly failing. Had the West achieved total success, however, the Soviets would have reduced the number of their troops in Europe by 11,500. Yet as that negotiation was closing down, Mikhail Gorbachev announced troop cuts of a half million—*forty-three times* the West's goal at the table. Of these, 240,000 are to be taken out of Europe, more than twenty times what we sought.

Arms Control Hype

Why me? Because I was fortunate enough to participate in three summits during nearly five years as Arms Control Director, from early 1983 until the INF Treaty was signed at the end of 1987. My position on the front lines merely reinforced my view that arms control is not what it's cracked up to be. . . .

Hyping arms control constitutes poor judgment. As President Carter's Secretary of Defense Harold Brown said truthfully, "Measured against these glittering possibilities, the achievements of arms negotiations to

Excerpted from Kenneth L. Adelman, *The Great Universal Embrace*. New York: Simon & Schuster, Inc., 1989. © 1989 by Kenneth L. Adelman. Reprinted by permission of Simon & Schuster, Inc.

date have been modest indeed. . . . In all, not much to show for thirty-five years of negotiations and twenty years of treaties."

Perhaps nowhere in life is the disparity greater between exalted public expectations and a dismal track record than in arms control. No area of science, medicine, or even public policy would continue to elicit so much hope after so many years of unsuccessful effort. Arms control must be approached as one of the intangibles of life, a rite seemingly needed to satisfy some deep longing in our collective soul.

The Armageddon Motif

Maybe the hopes we associate with arms control arise from the dread of annihilation that lies deep in what Carl Jung called our "collective unconscious." The Armageddon motif—that the earth will perish and mankind will end—interestingly predates the advent of nuclear weapons. It even predates the Christian notion of the Last Judgment and Armageddon itself. . . .

This archetype played out in the 1930s when annihilation literature focused on chemical weapons, especially combined with aerial warfare. British Prime Minister Neville Chamberlain was typical in portraying "people burrowing underground, trying to escape from poison gas, knowing that at any hour of the day or night death or mutilation was ready to come upon them." Total war would come about if any war erupted. "Whichever side may call itself the victor, there are no winners, but all are losers."

The ultimate irony was that, when the war in Europe came, the only weapon *not* used was chemical arms. This interwar fear-mongering turned out wrong, not because of any kindness by Hitler—he used gas to exterminate Jews—or of any international legal treaties—he gave them all short shrift—but simply because of deterrence. German intelligence estimated that the British and French had much larger stockpiles of chemical weapons than they actually had. The German High Command dreaded retaliation in kind if they initiated a chemical attack, and Hitler, who had been gassed as a young corporal in World War I, agreed.

"While nuclear arms catch the world's eyes, conventional arms cost the world's treasures in gold and blood."

After the war, the Armageddon motif of course fixed on nuclear weapons. This fostered a vast literature and veritable film festival on nuclear apocalypse. And it brought a wave of dire predictions, such as, in 1960 during relatively tranquil times, C.P. Snow prophesying that if events proceeded on their current path, nuclear war was "a certainty" within ten years.

When nuclear war became portrayed as our world's prime problem, nuclear arms control became its main solution. This perspective persists even though nuclear weapons have long been recognized as primarily political instruments rather than military weapons. They constitute what the great Chinese strategist Sun Tzu deemed the most effective of all instruments of war: "To subdue the enemy without fighting is the acme of skill.". . .

[The] dread of annihilation and resulting desire for arms control was restricted to U.S.-Soviet nuclear arms control. This is understandable, since herein lies the danger of total obliteration.

But it is also convenient. As Canadian Ambassador Alan Gottlieb once told me, arms control is the attempt to control *others'* arms, never one's own. For Britain, France, and China, what matters is controlling American and Soviet nuclear arms, not their own. They resist talks on their nuclear systems like the plague. And for the Soviets, arms control is naturally an attempt to control American arms. For West Germany and most other states, it is the effort to control the type of arms they lack, nuclear arms.

This is especially true of Third World countries, which have increased their military spending much faster than industrialized nations or Communist countries. Their ability to pay for arms is, however, far lower. Countries like Libya, Syria, Tanzania, and South Yemen spend more of their meager GNPs [gross national product] on arms than the average NATO [North Atlantic Treaty Organization] nation.

The Cost of Conventional War

While nuclear arms catch the world's eyes, conventional arms cost the world's treasures in gold and blood. Today, upward of 95 percent of all military spending goes for conventional arms. And it is with these, not nuclear bombs, that lives are lost. During this century, around 150,000 people perished by nuclear weapons while one thousand *times* that number, or 150 million, have been killed by conventional weapons. In one recent year, 1982, some thirty-six conventional wars raged, killing some four million people and wounding countless more.

So this century's Armageddon has been conventional, not nuclear. Even if fears of a nuclear Armageddon are legitimate, given the awesome power of nuclear weapons, then the presumed solution of arms control is misplaced. For as long as nuclear weapons exist, which they will forevermore, the potential for total annihilation will likewise exist. Arms control on short- or medium-range nuclear weapons, as in the INF Treaty, does not change that equation one iota. These systems constitute small change compared with the big strategic stuff.

But neither would strategic arms control have much of an impact on preventing what folks fear most—total destruction. The strategic reductions envisioned in START [Strategic Arms Reduction Talks] would still leave both sides with many times more strategic weapons than they had during the Cuban missile crisis, to say nothing of the thousands of tactical and battlefield systems and air- and sea-launched nuclear cruise missiles. In a nutshell, the nuclear numbers have become so astronomical that no arms control effort could ever preclude the possibility of Armageddon.

Even if arms control could reduce these numbers, having fewer nuclear weapons does not necessarily mean less risk of nuclear war. Indeed, having fewer nuclear weapons might actually raise the likelihood of war by prompting either side to "use 'em or lose 'em" for fear that its smaller nuclear force was at greater risk than before. Today's high numbers allow for a redundancy that helps preclude this type of panic in a crisis.

"START, conventional arms control, and a chemical weapons ban cannot deliver anything remote to what they now promise."

The proper goal of strategic arms control is not across-the-board reductions but selective reductions in those systems most vulnerable and those with a "first strike capacity." This is fine in theory but tough in practice. For instance, a 50 percent cut in strategic arms would naturally prompt each side to cut its oldest weapons while retaining its newest ones. But it is generally the newest systems that have the most warheads packed on the fewest missiles. Our MX and the Soviets' new SS-24 both have up to ten warheads apiece, while our Trident submarines have some 180 or so warheads.

The Main Problem

Rather than enter this strange and specialized realm of nuclear accountancy, we should stay on the grand plane: nuclear weapons per se are not the main problem. U.S.-Soviet competition in regional conflicts, stemming from Moscow's acts of or support for aggression, *is* the main problem. These grabs can always escalate and lead to large-scale, and then nuclear, conflict.

Thank goodness both Moscow and Washington are keenly aware of this. Hence the remarkable record, rather unique in the annals of history, of two major powers assiduously avoiding instances where their military forces rub up against the other. And hence the growing taboo against "nuclear gunboat diplomacy" in the postwar years. . . .

As President Reagan never tired of saying, nations do not develop mistrust because of arms. Rather, they

develop arms because of mistrust. Western mistrust has been based on the Soviets' seventy-year record of repression within and aggression beyond its borders. With Gorbachev's reforms, this may be changing; a less aggressive and repressive Soviet Union would do more to preclude [a] . . . nuclear nightmare than a hundred START treaties or a thousand INF accords. . . .

Return to Traditional Diplomacy

Mine is a call for a return to traditional diplomacy, though one conducted by military experts more steeped in security matters than traditional diplomats. Gone would be the formal, quasi-public discussions in Geneva over arcane strategic matters. Back would be a focus on the big picture of greater security with fewer nuclear weapons.

Replacing the elaborate arms control ritual would be what in March of 1933 Winston Churchill called "private interchanges." He illustrated the dialogue as "'If you will not do this, we shall not have to do that,' and 'If your program did not start so early, ours would begin even later.'" Churchill believed that

> a greater advance and progress towards a diminution of expenditure on armaments might have been achieved by these methods than by the conferences and schemes of disarmament which have been put forward at Geneva.

He was right then and would be right now. For here too is an approach that has been tried and that has largely succeeded, as we will see.

The case for this approach of "arms control without agreements" rests on three propositions. First, the existing way will not work. Second, this way has, can, and is likely to work. Third, it is inevitable anyway.

First, the bad news: START, conventional arms control, and a chemical weapons ban cannot deliver anything remote to what they now promise. They may deliver nothing good at all.

No arms agreement in these areas can be comprehensive. So each opens itself up to the "balloon argument," namely that military capabilities pushed down by an arms accord in one area will pop up in others. Reductions in tanks and artillery will heighten competition in air-delivered weapons; cuts in air-launched cruise missiles will cause expansion in sea-launched cruise missiles or in bomber capabilities; and so forth.

No accord in these realms can be well verified, either. The strategic weapons now coming on-stream —cruise missiles and mobile land-based missiles—are exceedingly tough to verify. In June 1989, Chairman of the Joint Chiefs William J. Crowe Jr. said the "verification questions that remain on START" are "like horsemeat—the more you chew them, the bigger they get."

Contrary to public expectation, the ground-breaking verification provisions in the INF Treaty still fall short of sound verification for START. A treaty totally banning all chemical weapons defies *any* verification;

such substances can be made in fertilizer plants and transported in perfectly harmless vats to become deadly only when mixed with other harmless substances.

As for a conventional arms accord, measuring the precise numbers of troops or tanks is daunting. Small weapons are tough to verify, as we learned from INF. Despite our intelligence focus here since 1982, we still erred badly. For we had no idea the Soviets deployed *any* SS-23 missiles in East Germany, until they told us so. We had long figured a total of 24 to 36 of these systems deployed anywhere, whereas they actually had 167.

Fortunately, the Bush administration staved off new negotiations on short-range nuclear weapons, for a while at least. During NATO's gala 40th birthday summit in Brussels in May 1989, this contentious issue was virtually resolved, even though no leader would openly admit it. The Americans and British had wanted modernization but no negotiations; the Germans and Russians wanted negotiations but no modernization; the outcome will be *neither* negotiations *nor* modernization.

This upshot was practically inevitable. For the only country that could conduct such negotiations, the U.S., didn't want them. And the only country which could deploy such modernized systems, West Germany, didn't want *them*. So neither will have either.

Yet that summit raised false hopes for vast conventional arms control in Europe. President Bush dazzled the NATO gathering with a spanking new proposal, pushing up the completion date for such an accord while heaping upon it yet more obstacles. Bush apparently hopes to end forty years of Soviet conventional superiority via a treaty drafted and agreed to in six to twelve months' time, although the full Western proposal would not be presented to the Soviets for several of those months. . . .

No Greater Stability

It is unlikely that any of these arms negotiations will result in greater stability. Sophisticated arguments can be (and have been) made that our security would be hurt, not helped, by the START proposal *we* offered late in the Reagan administration, let alone the one that could come out of Geneva after much haggling. President Bush's National Security Advisor, Brent Scowcroft, joined experts John Deutch and R. James Woolsey in concluding that START could push us to "a new kind of triad: vulnerability, wishful thinking, and a hair trigger." Not terribly consoling words, but true.

And none of these would necessarily, at the end of the day, save money. Here the argument is more tentative since the particulars become more critical. Still, logic and history dampen hopes for any "peace dividend" arising from arms control. . . .

What about the rest? Where does my rather dismal assessment on other formal arms control efforts leave us? In a promising situation, oddly enough, wherein we can adopt a more direct approach to achieve our goals. Friedrich Nietzsche once observed that the most common form of human stupidity is forgetting what one is trying to do.

What we have been trying to do in arms control, rather than simply conclude a treaty, is to reduce the risks of war, especially nuclear war; lower the number of destabilizing weapons, especially nuclear weapons; reduce costs; improve warning time and intelligence gathering; and reduce collateral damage if, God forbid, deterrence should ever fail.

Yet these goals have been overshadowed by the lunge for an agreement. No rational being would ask a senator, "Are you going to conclude a bill this year?" without being queried in turn "What kind of bill? To accomplish what?" Yet rational people constantly ask a president, "Are you going to conclude an arms agreement this year?" while showing scant concern about the contents or repercussions of an agreement.

"Rather than go for an agreement without real arms control, we should seek real arms control without necessarily going for an agreement."

The public has been led to believe that concluding an accord that lowers virtually *none* of the Soviet threat is still to be applauded; witness SALT [Strategic Arms Limitation Talks] II, touted as among President Carter's greatest accomplishments. A faulty agreement is thus prized more than having no agreement.

The arms control hype has gotten completely out of hand. This was not Ronald Reagan's doing, though he did nothing to dampen it and much to heighten it. Reagan followed in a rich tradition of presidents from John F. Kennedy, who got nuclear arms control off to a roaring start by extolling the 1963 Limited Test Ban Treaty as a key step in "man's effort to escape from the darkening prospects of more destruction." These soaring words for an environmental accord that resulted in more nuclear bomb tests than before!

Such soaring words, said so often. No wonder arms control became the centerpiece of U.S.-Soviet relations. As always, my predecessor at ACDA [Arms Control and Disarmament Agency], Paul Warnke, said it clearest: "I think that strategic arms control is the bellwether of U.S.-Soviet relations. If we can't agree not to blow one another up and probably obliterate the rest of the world with us, then how can we agree on anything else?" What wild words, since neither side wished to blow up the other, themselves, or the world even before strategic arms control came along. And neither side has wanted to since. Yet neither would be prevented from doing so because of any arms accord.

As told, at times an arms agreement, or even the pursuit of one, has moved us *farther* from our goals. It is certain that the strategic balance is less favorable to us today than when strategic talks opened in 1969. It is contentious (but I believe true) that strategic stability is lower today than it would have been absent the arms control process. Had it not been for arms control, for instance, we would not have ended up with fifty highly capable, highly warheaded, highly vulnerable MX missiles deployed in the worst manner imaginable.

If my premises are correct—that traditional U.S.-Soviet arms control is conceptually bankrupt and practically blocked; that sound verifiable accords on strategic, conventional, or chemical weapons cannot even be written up by the Americans, let alone negotiated out with the Soviets—we should face the music and change our act. Rather than go for an agreement without real arms control, we should seek real arms control without necessarily going for an agreement.

We should adopt an approach of reciprocal though individual moves. The United States and the Soviet Union would thereby design programs to enhance their own security and then discuss those policies with each other, ideally coordinating them. Adopting this less formal approach keeps top officials from being lost among the trees and helps them see the forest: how to reduce the risks of war, the number of nuclear weapons, and so forth. They could more clearly see and build strategic systems that would be most survivable, sensible, and affordable rather than those that would be easiest for the other side to verify or mightiest to use as a bargaining chip in Geneva. . . .

The End of the Cold War?

For forty years, the West has been waiting breathlessly for an end to the Cold War. According to one scholar, Dr. Burton Marshall, top U.S. officials or opinion molders have declared a turning point in U.S.-Soviet relations on fifty-four different occasions in recent decades. None of them turned out to be definitive or lasting. But the fifty-fifth may be.

Sure, the possibility still exists that Moscow might lash out while communism gasps its final breath. Or Gorbachev could take Henry IV's deathbed advice to his son Henry V and declare war so as to "busy giddy minds with foreign quarrels." But I doubt either will happen. While less predictable today than ever before, the Bear is just too sick, and Gorbachev is just too practical. He too can see that communism is on its deathbed and communist states everywhere are on their backs. The high-water mark of the Soviet empire passed in the 1970s; what lies ahead is shallows and miseries.

Since Americans heralded the demise of the Cold War on fifty-four occasions when there was scant

cause, it is not hard to understand their anticipation today when there is good cause. It may be difficult for our government to manage the devolution of the Soviet empire in a sensible way. We can take consolation, however, in the knowledge that it will be far tougher for them to do so. . . .

We should hold back the hosannas until we know we *have* won the Cold War, which may not be knowable (or even happen) for a good number of years. We should realize that there will surely be bumps and diversions along the way.

Meanwhile, we must cope with runaway Western sentiment that the Soviet threat is passing, if not passé. Support for increased Western military spending has evaporated and will continue to evaporate unless Gorbachev resorts to some outlandish action, à la another Afghanistan, which is most improbable. Hence Western political will is lacking to support more missiles or new and terribly pricey strategic bombers. Strategic defense, executed deliberately and coolly, can gain support, but only if (as I expect) SDI [Strategic Defense Initiative] research pans out and is accompanied by cuts in other strategic programs.

Defense Spending Declining

In a nutshell, real arms reductions are coming. They will result not from negotiations in Geneva, but from deliberations in Washington and Moscow. Budget cutters will do what arms controllers have generally failed to do. In fact, they are already doing so, with U.S. defense spending declining by more than 13 percent in real terms between 1985 and 1989 and announced Soviet cuts of the same magnitude for 1990. We should now let nature take its course, watch the improvement of strategic stability and the draw-down of conventional forces and nuclear weaponry outside of what happens around the tables in Vienna and Geneva.

"Real arms reductions are coming."

Why, one may reasonably ask, would Gorbachev reduce his conventional forces (for example) unilaterally rather than multilaterally in the Vienna talks, where he can get something from NATO in return? Simply because he *knows* that he will get something from NATO in return for more *unilateral* cuts. No one glancing at the mood across Western Europe today can miss the tremendous benefits he's already reaping from the limited conventional cuts announced thus far.

Besides, such a process is as natural as the formal arms control process is artificial. When two nations move from being fierce enemies to strong adversaries to mere rivals, they commonly lower their armaments—not through arms control, but through downgrading the threat. This has happened repeatedly in history—with the United States and Canada in mid-nineteenth century, the British and the French in the early years of this century, and China and India during the fifteen years after their 1962 conflict. It is happening now between Russia and China.

If, as I contend, formal arms control has little to contribute to this quasi-inevitable process, at least it can step aside and let the goals of arms control be realized without the formal process blocking their realization. . . .

Future Views of Arms Control

After some time, historians can and surely will look back to ponder what caused the arms control hype of our era. Surely hope lay behind it, a hope that later may be transferred to SDI and to the obvious warming of relations between the Soviet government and ours (which can only follow the warming of relations between the Soviet government and its own people).

"Surely historians will see that tension and antagonism lay behind the arms control hype."

And surely historians will see that tension and antagonism lay behind the arms control hype. When the U.S.-Soviet relationship was strictly antagonistic, formal arms control came to be regarded as its barometer. It was treasured, not for what it accomplished in the way of its goals, but for what it indicated in the way of U.S.-Soviet relations.

Once the superpower relationship turns less antagonistic, as is happening now, formal arms control will lose even this utility. So, in perhaps the ultimate irony, future historians will come to view formal arms control as just another accoutrement of the Cold War.

Kenneth L. Adelman was United States arms control director from 1983 to 1988 and Deputy United States Representative to the United Nations from 1981 to 1983. Now a nationally syndicated columnist, Adelman is also vice president of the Institute for Contemporary Studies in San Francisco. He teaches at Georgetown University in Washington, D.C., and at Johns Hopkins School of Advanced International Studies in Baltimore.

"To hold START hostage to a conventional-arms-control agreement is tantamount to postponing it indefinitely."

The U.S. Should Negotiate a START Agreement

Robert S. McNamara

The new Soviet leadership has given arms-reduction efforts the highest possible priority. But here in the United States the political consensus that led to successful ratification of the INF [intermediate-range nuclear forces] Treaty and support for START [Strategic Arms Reduction Talks] is severely threatened. I fear that the unresolved issues in the START Treaty and the mounting criticism of the arms-control negotiations, coming from many quarters, could cause this consensus to unravel.

A Boldness of Vision

Gorbachev has undeniably shown a boldness of vision, advocating sweeping measures to reverse the nuclear-arms race and reduce conventional arms in Europe. These bold proposals have been accompanied by remarkable flexibility at the negotiating table. The Soviets agreed to unequal reductions and extensive on-site inspection in the INF Treaty. And the framework of the START Treaty, calling for deep reductions in strategic arms, which the Soviets have agreed to, largely reflects U.S. concerns about Soviet advantages in ballistic missiles. In conventional arms, Mikhail Gorbachev has not only proposed a three-phase plan for substantial cuts, but on December 7, 1988, announced major unilateral reductions.

Not long ago, in the days when Soviet officials were always rejecting U.S. proposals, Gorbachev's approach to diplomacy would have been unimaginable. Yet, now that the Soviet Union has a leader who, in the words of British Prime Minister Margaret Thatcher, is someone the West "can do business with," a great debate is brewing among experts here in the United States about whether and how to pursue arms reductions.

The U.S. opponents of START have developed a three-pronged attack. One group argues that pursuing

Excerpted from Robert S. McNamara, *Out of the Cold.* New York: Simon & Schuster, Inc., 1989. © 1989 by Robert S. McNamara. Reprinted by permission of Simon & Schuster, Inc.

START is dangerous because it will inevitably involve limiting our right to test and deploy strategic defenses. Others claim that reductions in strategic arms will erode the credibility of "extended deterrence," thereby increasing the chances of conventional aggression by the Soviets. A third group believes that START's specific reductions will leave U.S. nuclear forces more vulnerable than they are today.

Since the frequency and intensity of these criticisms may well grow as the negotiations continue, they merit careful examination.

The most important criticism of START relates to its effect on crisis stability. This is a valid—indeed, central—issue. The principal aim of arms control is not simply to reduce the number of weapons but to make nuclear war less likely by reducing incentives for either country to launch a preemptive attack in a crisis. For the United States, this goal—of enhancing crisis stability—means ensuring the survivability of U.S. systems by curtailing the capability of Soviet forces and by encouraging a Soviet shift to a more stabilizing force posture.

Decrease in Submarines

The critics of START state it will increase the vulnerability of U.S. land- and sea-based systems. They point out that the number of submarines will decline from the present thirty-five to some sixteen to eighteen. The critics' mistake is to equate reduced numbers with reduced survivability. With or without START, it is planned to sharply decrease the number of submarines. We have decided that the increased range, quietness and efficiency of Trident as compared to Poseidon submarines justifies concentrating submarine-launched-ballistic-missile (SLBM) warheads in a smaller number of boats. There are simply no breakthroughs in antisubmarine warfare on the horizon that would make our Trident submarines vulnerable. Over time, if Soviet capabilities in this area were to improve dramatically, we are free under START to increase the number of submarines we deploy by reducing the number of weapons each one carries.

The issue that seems to generate a disproportionate share of controversy is the question of ICBM [intercontinental ballistic missiles] vulnerability. Some critics suggest START will make our fixed ICBMs more vulnerable because the number of U.S. silos may be reduced to a greater extent than the number of Soviet SS-18 warheads. Thus, critics claim that START will threaten crisis stability, since the ratio of Soviet counterforce warheads to fixed U.S. ICBM targets would, in that case, rise.

This need not occur. The Scowcroft Commission, appointed by President Ronald Reagan, concluded, in a judgment which I share, that our current triad of land-, sea- and air-based strategic nuclear forces is invulnerable in the face of current and prospective Soviet forces. While our land-based missiles may become somewhat more vulnerable as Soviet systems are further modernized, START need not make the situation worse. On the contrary, unless we choose to deploy our forces under START in a greatly reduced number of fixed—as opposed to mobile—launchers, the critics' fears will not be realized. We can make the decision to do otherwise. Consequently, depending on the mix we choose of existing and modified Minuteman missiles, START can modestly or greatly diminish whatever Soviet incentives may exist to attack U.S. ICBMs.

START and Extended Deterrence

Perhaps the most troublesome of all the attacks on START is the one which seeks to make a connection between strategic-arms reductions and the likelihood of conventional war in Europe. Those who express such concerns propose that START be formally linked to improvements in the conventional balance. They claim that START would reduce the deterrent capability of our strategic nuclear forces. But I see no possible way in which this could occur. Because the reductions in START are so balanced and will enhance the overall survivability of U.S. strategic forces, and because the United States would still retain nuclear weapons numerous enough and flexible enough to support NATO [North Atlantic Treaty Organization] strategy, our capability to use nuclear forces in defense of Europe would remain unchanged. Therefore, whatever role strategic nuclear forces now play in deterring the threat of Soviet conventional aggression—and I regard that role as minimal—they would play an equal or greater role after they are adjusted to the treaty limits.

To hold START hostage to a conventional-arms-control agreement is tantamount to postponing it indefinitely. While the opportunity to achieve conventional-arms reductions is, I believe, greater today than ever before, the extreme complexity of the subject matter makes it abundantly clear that actual progress will take many years.

A major unresolved issue standing in the way of agreement on START is the status of the Strategic Defense Initiative (SDI), the anti-ballistic missile system proposed by President Reagan. My opposition to this program is well known. I believe it poses grave risks far out of line with any possible benefits.

Those risks include: its enormous waste of resources, which, depending on the choice of system, may be measured in the hundreds of billions of dollars; the fact that it will initiate an arms race in outer space and fuel the competition in strategic arms on earth; and the possibility that it would increase the temptation for preemptive attack in a period of crisis.

A Grim Fact of Life

Meanwhile, the potential benefits of SDI are minimal. Given the enormous destructive power of nuclear weapons, no defense in the foreseeable future, no matter how extensive and costly, can protect the population of the United States. This vulnerability is not a policy choice but a grim fact of life in the nuclear age.

Despite the grave risks and minimal benefits, proponents of the SDI insist on keeping alive its test and development program, which they hope and expect will lead to deployment of the system in the mid-1990s. They are unwilling to recognize that neither the Soviets nor the United States will or should accept a limitation on strategic offensive weapons if the opponent is permitted an unlimited defense. Former Secretary of Defense Caspar Weinberger has gone so far as to propose that the United States withdraw from the ABM [anti-ballistic missiles] Treaty. And yet the inextricable link between reductions in strategic offensive arms and restrictions on strategic defenses remains as true today as it was in the late 1960s and early 1970s, when the ABM Treaty was first formulated. I know that the Joint Chiefs of Staff who advised me in the 1960s shared this view. I would be very surprised if the current Joint Chiefs do not still share it. If the Chiefs were pressed whether they could support deep reductions in U.S. strategic forces, such as those START would require, combined with an unlimited Soviet deployment of a nationwide defense, I can't imagine they would say yes.

"START can modestly or greatly diminish whatever Soviet incentives may exist to attack U.S. ICBMs."

The U.S. and Soviet governments must find a way to finesse this impasse over missile defense. If the United States insists that strategic defense be given free rein, we cannot expect the Soviet Union to implement arms reductions, nor should we in the United States be willing to do so. The best solution would be for the new Administration to endorse the traditional interpretation of the ABM Treaty, to heed the bipartisan advice offered by six previous Secretaries of Defense—three Republicans, Melvin R. Laird, Elliot L. Richardson and James R. Schlesinger, and three Democrats, Clark M. Clifford, Harold Brown and myself. We stated that limiting the testing and deployment of both U.S. and Soviet strategic defenses through the ABM Treaty is critical to U.S. security and "makes possible negotiation of substantial reductions in strategic offensive forces."

Alternatively, the impasse could be overcome through Soviet action. Recognizing that Congress has not permitted the President to conduct SDI tests that go beyond the traditional interpretation of the ABM Treaty, the Soviets could make a unilateral statement to the effect that their national interests would be jeopardized by U.S. actions inconsistent with the ABM Treaty. The right of a nation to alter its legal obligations under a treaty like START because of a threat to its national interest is well established. The United States would be hard pressed to justify opposition to such an expression of sovereignty.

Deep Reductions

Those who oppose START because it will involve limits on strategic defenses are correct in their assumption, but incorrect in their conclusion. Obtaining deep reductions in Soviet ballistic-missile forces in exchange for restrictions on a program of such dubious value and potentially dangerous consequences is an offer we should be delighted to accept.

I do not believe that any leg of this "triad of opposition" to START will stand up under careful scrutiny. However, if these attacks continue, they will certainly be divisive. They could yet prove to be decisive, and this historic opportunity would be missed or substantially deferred. . . .

The "short-term" program—completing the START negotiations, achieving a balance of conventional forces in Europe, reducing the number of tactical nuclear weapons and strengthening confidence-building measures—will greatly improve crisis stability. However, after it is completed, NATO and the Warsaw Pact will retain thousands of nuclear warheads, and NATO's strategy will continue to be based on first use of these weapons under certain circumstances. The danger of nuclear war—the risk of destruction of our society—will have been reduced but not eliminated. Can we go further? Surely the answer must be yes.

More and more political and military leaders are accepting that major changes in NATO's nuclear strategy are required. Some are going so far as to state that our long-term objective should be to return, insofar as practical, to a nonnuclear world. Illustrative of the changes in thinking which are now under way

was this statement by the former Chancellor of the Federal Republic of Germany, Helmut Schmidt, in 1987:

> Flexible response [NATO's current strategy calling for the use of nuclear weapons] is nonsense. Not out of date but nonsense, because it puts at risk the lives of sixty million Germans and some fifteen million Dutch and I don't know how many million Belgian people and others who live on Continental European soil. The Western idea—which was created in the 1950s—that we should be willing to use nuclear weapons first, in order to make up for our so-called conventional deficiency, has never convinced me. I can assure you that after the use of nuclear weapons on German soil, the war would be over as far as the Germans go, because they would just throw up their hands. To fight on after nuclear destruction of your own nation has started is a very unlikely scenario, and you need to be a mathematician or a military brain to believe such nonsense.

Similar views have been expressed by a number of NATO military experts:

Field Marshal Lord Carver, Chief of the British Defense Staff from 1973 to 1976, is totally opposed to NATO's ever initiating the use of nuclear weapons. He has said, "At the theater or tactical level any nuclear exchange, however limited it might be, is bound to leave NATO worse off in comparison to the Warsaw Pact, in terms both of military and civilian casualties and destruction."

General Johannes Steinhoff, the former Luftwaffe Chief of Staff, stated, "I am firmly opposed to the tactical use [of nuclear weapons] on our soil."

Admiral Noel A. Gayler, former Commander in Chief of U.S. ground, air and sea forces in the Pacific, wrote: "There is no sensible military use of any of our nuclear forces."

Melvin Laird, Secretary of Defense in the Nixon Administration, shares the views of the military commanders. He said, "These weapons are useless for military purposes."

Such diverse groups and individuals as the U.S. Catholic bishops, Henry Kissinger and antinuclear advocates have all stated that nuclear deterrence is an untenable strategy which the U.S. public will not support indefinitely.

"There is no sensible military use of any of our nuclear forces."

And President Reagan repeatedly asserted that a nuclear war cannot be won and must never be fought—assertions which clearly imply disavowal of NATO's current strategy of flexible response. His proposal for the Strategic Defense Initiative went even further and had as its objective the complete elimination of all nuclear weapons.

In spite of such statements, many NATO security

experts—I would say most—have not yet been willing to support a basic change in NATO's nuclear strategy. For example, with reference to a proposal for eliminating nuclear weapons, Zbigniew Brzezinski, President Jimmy Carter's national-security adviser, said, "It is a plan for making the world safe for conventional warfare. I am therefore not enthusiastic about it."

But two developments in the years ahead are likely to lead to a shift in such judgments: Restructuring of conventional forces in Europe into defensive postures, at levels that are clearly in balance, will sharply reduce the justification for the nuclear-deterrent force. And arms negotiations which result in nuclear forces of equal capabilities—forces with which neither side could win a nuclear war—will remove the deterrent capability of the force. Such "no-win" strategic-force structures are implicit in the proposed START agreements.

If it were decided to move away from nuclear deterrence, how would this be done?

Total Elimination

Mikhail Gorbachev has proposed that the United States and the Soviet Union aim at achieving the total elimination of all nuclear weapons by the year 2000. But the genie is out of the bottle—we cannot remove from men's minds the knowledge of how to build nuclear warheads. Therefore, unless technologies and procedures can be developed to ensure detection of any steps toward building a single nuclear bomb by any nation or terrorist group—and such safeguards are not on the horizon—an agreement for total nuclear disarmament will almost certainly degenerate into an unstable rearmament race. Thus, despite the desirability of a world without nuclear weapons, an agreement to that end does not appear feasible either today or for the foreseeable future.

"Policing an arms agreement that restricted each side to a small number of warheads is quite feasible with present verification technology."

However, if NATO, the Warsaw Pact and the other nuclear powers were to agree, in principle, that each nation's nuclear force would be no larger than was needed to deter cheating, how large might such a force be? Policing an arms agreement that restricted each side to a small number of warheads is quite feasible with present verification technology. The number of warheads required for a force sufficiently large to deter cheating would be determined by the number any nation could build without detection. I know of no studies which point to what that number might be—they should be initiated—but surely it would not exceed a few hundred, say at most five hundred. Very possibly it would be far less, perhaps in the tens.

Such an agreement could be achieved only over a period of years—say by the year 2000—but should we not set it as our ultimate objective and lay out a series of steps to move toward it?

The Budgetary Impact

As we progress through the arms control agenda, the U.S. defense budget could be reduced substantially. It might well be possible, within six to eight years, to cut military expenditures in half in relation to GNP [gross national product]—i.e., to 3 percent. That would make available, in 1989 dollars and in relation to 1989 GNP, $150 billion per year to be divided between additional personal consumption and the financing of the pressing human and physical infrastructure needs of both our own and Third World societies. It would go far toward assuring that U.S. industry would be in the lead in the technological world of the twenty-first century. Lest it be thought a U.S. defense budget of 3 percent of GNP is a fantasy, we should remember that today Japan's defense expenditures are 1 percent of GNP, Canada's 3 percent, and the average of all NATO nations', excluding the United States, 3 percent.

The political actions and arms-control proposals which I have outlined . . . should be supplemented by measures of economic and environmental cooperation and by scientific and cultural exchanges. All would be designed to integrate the Soviet Union more fully into our increasingly interdependent global order. Each action could be taken independently of the others. But together they would exert a powerful synergistic effect which would indeed make the whole greater than the sum of its parts.

Robert S. McNamara was Secretary of Defense under Presidents John F. Kennedy and Lyndon B. Johnson. Formerly president of the World Bank and of Ford Motor Company, McNamara is a prolific author. His books include Blundering into Disaster: Surviving the First Century of the Nuclear Age, The Future Role of the World Bank, *and* The Essence of Security.

"Conventional forces reductions now should have higher priority than the Strategic Arms Reduction Talks."

The U.S. Should Negotiate a Conventional Arms Agreement

Jay P. Kosminsky

Secretary of State James Baker heads for Moscow for talks with his counterpart, Soviet Foreign Minister Eduard Shevardnadze. At the top of their agenda is to determine how the improved climate of East-West relations will affect U.S. and Soviet arsenals. What Baker should tell his host is that the United States is ready to move quickly on reducing conventional forces but is determined to proceed more cautiously on nuclear weapons.

This, of course, was the message of George Bush's State of the Union speech, when he said "the Soviet military threat in Europe is diminishing but we see little change in Soviet strategic modernization." Bush then added substance to this general assessment by proposing deep cuts in Soviet and U.S. forces in Europe and by saying that "the time is right to move forward on a conventional arms control agreement." He was referring to the Conventional Forces in Europe (CFE) negotiations underway in Vienna. Bush's new plan would cut American and Soviet forces in central Europe to at most 195,000. The U.S. now has close to 300,000 troops in this region, and Moscow close to 600,000. Bush's proposed reductions ultimately would save Americans close to $10 billion per year if eliminated forces are demobilized.

Reducing a Military Threat

Conventional forces reductions now should have higher priority than the Strategic Arms Reduction Talks, known as START, for a number of reasons. For one thing, although Bush and Soviet leader Mikhail Gorbachev decided at their December 1989 Malta summit to set a June 1990 negotiating deadline for a START treaty, tough issues remain. Even some of Bush's top arms control advisors are saying that a solid START treaty will take at least another year and a half to work out. For another thing, only CFE can

help eliminate the greatest remaining military threat to NATO [North Atlantic Treaty Organization] and to the survival of the European Revolutions of 1989— the heavy concentration of Soviet forces in the heart of Europe.

Meanwhile, another issue emerged that will demand Baker's and Shevardnadze's prime time: the sudden and massive communist offensive, backed by Soviet weapons and helped by Soviet advisors, in Angola. Ostensibly, Gorbachev's "new thinking" in foreign affairs is supposed to end meddling in Angola and other regional conflicts. Baker thus should ask the Russians: "If you don't keep your word on Angola, how can you be trusted on arms reduction matters?"

START talks, of course, are important. The point is, in fact, they are too important to be rushed. This would yield a sloppy treaty which then would encounter fierce opposition in the Senate. Key problems remain on START. Among them:

Procedures for verifying a START accord will be particularly difficult to work out since the agreement is likely to allow mobile missiles, like the Soviet SS-24 and SS-25 and proposed U.S. *Midgetman*. These are more difficult to find and count than missiles stationed in fixed silos. Procedures still must be worked out also for counting and monitoring how many missiles are manufactured in Soviet and American factories. If the U.S. rushes for an agreement on verification by the June START deadline, the treaty could contain loopholes and ambiguities that Moscow could exploit.

Common Understanding

Soviet statements on whether the U.S. Strategic Defense Initiative (SDI) does or does not impede a START agreement are ambiguous and even contradictory. At his September 1989 meeting with Baker in Wyoming, for example, Shevardnadze said that Moscow no longer would insist on a separate treaty limiting SDI as a precondition for a START

Jay P. Kosminsky, "Bush Should Move Full Speed on Conventional Arms Limits, Slow on Nuclear," The Heritage Foundation *Backgrounder*, February 5, 1990. Reprinted with permission.

Treaty. Then he turned around and said that Moscow would insist on a "common understanding" allowing the Soviet Union to repudiate a START Treaty if the U.S. moved to deploy SDI. The Bush Administration wisely refuses to accept Moscow's new position. Bush told Congress in his State of the Union address that "we must sustain . . . the Strategic Defense Initiative."

While Moscow officially has dropped its demand to count SLCMs [submarine-launched cruise missiles] under START limits, it is insisting on a separate agreement to ban SLCMs, which it says must be signed along with START. This Washington opposes, primarily because it does not believe that SLCMs adequately can be verified.

The two sides still remain far apart on how many mobile missiles and mobile missile warheads will be allowed under a START treaty, and on procedures for verifying how many mobile missiles each side has. To further complicate matters, it was reported on January 15, 1990, that National Security Advisor Brent Scowcroft is trying to change the U.S. position on mobile missiles to allow only single-warhead mobile missiles like the Soviet SS-25 or U.S. *Midgetman*, while banning those with multiple-warheads including the Soviet SS-24 and U.S. rail-based MX.

Other unresolved START issues include how to count such planes as the B-2 "Stealth" bomber, designed to carry about sixteen nuclear bombs, and how to count planes equipped with Air Launched Cruise Missiles (ALCMs), which can be launched outside Soviet territory at targets in the Soviet Union. The two sides also have not worked out the details of how to define "new types" of so-called "heavy" missiles like the Soviet SS-18, which can carry huge payloads, or the issue of what information on missile and warhead performance each side can "encrypt," or code, during missile tests. Given time, these issues probably can be worked out to the satisfaction of the American people and the Senate, which must ratify a START treaty. However, if Bush tries to resolve them all under the pressures of a self-imposed negotiating deadline, he is liable to make costly mistakes.

"A good CFE agreement will enable the West to monitor and verify Soviet troop reductions."

Unlike the START negotiations, which have been bogged down over many of the same issues for years, CFE negotiations have been progressing at a speed unprecedented in arms control history. It is important to maintain this momentum. By staying ahead of the tumultuous events in Europe, CFE can ensure that these events help drive Soviet forces from Eastern Europe and back to Soviet territory.

When negotiations got underway in March 1989 in Vienna, CFE looked like the West's only chance of eliminating Soviet military advantages over NATO in Europe in tanks, armored combat vehicles, artillery, and aircraft, and of forcing the Soviet Union to withdraw large numbers of forces from Eastern Europe. Since then, however, the revolution sweeping through Eastern Europe promises to push back Soviet troops altogether without the aid of arms control. Already Czechoslovakia, Hungary, and Poland have asked Moscow to take its forces off their territory. And surely a freely-elected East German government would follow suit.

Framework for Withdrawal

Change driven by CFE is better than allowing events in Europe to run their course. A good CFE agreement will enable the West to monitor and verify Soviet troop reductions and the destruction of roughly 100,000 major Soviet weapons; it can "lock in" Soviet reductions as a hedge against a possible reversal of policy in Moscow; and it will provide a framework for the orderly withdrawal of tens or perhaps hundreds of thousands of U.S. forces from Europe.

Bush's new proposal thus moves CFE in the right direction. It would set U.S. and Soviet troop levels in Central Europe—basically in Germany and most of the East European countries—at 195,000; exempted from this ceiling are the 30,000 U.S. troops in Britain, Italy, and Turkey. This tells Moscow that the U.S. reserves the right to continue stationing some, limited, forces in Europe as long as NATO allies want them there and even if Moscow withdraws its troops to within its own territory. Said Bush in his State of the Union Address: "I agree with our European allies that an American military presence in Europe is essential—and that it should not be tied solely to the Soviet military presence in Eastern Europe."

The one major problem with the Bush proposal, and with CFE in general, is that limits set on Soviet forces in Eastern Europe could be construed as implicit recognition that these forces have a right to be stationed there. As Bush moves forward on CFE, he will have to dispel this notion by supporting the sovereign rights of all European countries to determine whose troops can and can not be stationed on their territory. In this context, Bush should draw a clear distinction between the status of U.S. and Soviet forces in Europe. U.S. forces legitimately are in Europe at the invitation of democratically elected allied governments. Soviet forces are in Europe at the invitation of now-deposed and discredited Soviet-backed regimes. They have no legitimacy and will gain none through CFE as long as the U.S. and its allies make this case strongly.

Instead of spending his time in Moscow trying to meet an impossible START deadline, Baker should follow up on the President's promising new CFE proposal, ensuring that CFE continues to be a force for rapid change in Europe. Baker should carry the following arms control messages to Shevardnadze:

- The U.S. is lifting its June START deadline. A START Treaty concluded by June 1990 is sure to contain loopholes that Moscow could exploit and which the Senate surely will expose during the ratification process. There is no reason to rush for a bad START agreement in June when an acceptable agreement might be had in 1991.

- Bush's ceilings on American and Soviet forces in Europe in no way sanctions the presence of Soviet forces in Eastern Europe against the will of East Europeans. The U.S. should insist on a CFE Treaty preamble that includes a declaration that the treaty imposes no obligation on any country to accept the stationing of foreign forces on its territory against its will. This will put America squarely on the side of Czechs, Hungarians, and Poles who through their governments have demanded the withdrawal of all Soviet forces from their territories.

"There is no reason to rush for a bad START agreement in June when an acceptable agreement might be had in 1991."

- The U.S. is open to cuts in U.S. and Soviet forces in Central Europe even deeper than those proposed by Bush so far. Pentagon and State Department officials privately report that Moscow is talking about cutting American and Soviet forces in Europe as low as 150,000. As long as Moscow is willing to accept Bush's distinction between U.S. forces in the central region and its forces on the periphery in Europe, Bush should be open to these deeper cuts.

Historical Change

Historical change in Europe is running in favor of the West, sweeping communist regimes and Soviet troops out of Europe in its wake. Through CFE the U.S. and its allies have an opportunity to help remove the major military threat to NATO and to the survival of the European Revolutions of 1989—Russian forces in Europe. CFE can serve the cause of change if Baker moves quickly, seeking a treaty that cuts Soviet forces deeply and enshrines the right of Czechs, Hungarians, Poles, Germans, and all Europeans to decide whose forces will or will not be stationed on their territory. For the time being, START should take a back seat to these objectives.

Jay P. Kosminsky is the deputy director of Defense Policy Studies and a national security policy analyst for The Heritage Foundation, a conservative think tank in Washington, D.C.

"It is time that the arms control process be extended to the superpower navies before an opportunity is lost to wind down the risks of the Cold War at sea."

Negotiating Cuts in Naval Weapons Would Promote U.S. Interests

Michael L. Ross

The rise of Mikhail Gorbachev and the subsequent improvement in East-West ties have led to many changes in the armed forces of the United States and the Soviet Union, but the restructuring of the superpower navies and the decline of the arms race at sea have attracted little notice. In the early 1980s, the high seas became the most dangerous battleground of the second Cold War. The U.S. and Soviet navies both undertook large shipbuilding programs devised to outflank each other. Each navy rose to greater prominence in its nation's military strategy, and each conducted daunting maritime exercises and deployed new nuclear weapons in parallel bids to project power globally.

With the ebbing of this Cold War, both navies now find themselves stranded in receding waters. Each has been forced to change its leadership, trim its budget, curtail operations overseas, and re-evaluate its fundamental purposes. Yet both navies remain uniquely exempt from the arms control negotiations that will reshape the land and air forces of the two superpowers. Until the two governments agree to limits on their naval forces, each will continue to build and deploy naval weapons and vessels that the other finds threatening. More important, unless the United States reconsiders its opposition to naval arms control, it will miss a critical opportunity to bring stability to the high seas and eliminate a troublesome category of nuclear weapons: tactical nuclear weapons. . . .

Early Retirement

In April 1989 *The New York Times* reported that the U.S. Navy was quietly retiring ahead of schedule three aging nuclear weapons systems: the ASROC ship-launched antisubmarine rocket, the SUBROC submarine-launched antisubmarine rocket, and the

Michael L. Ross, "Disarmament at Sea." Reprinted with permission from FOREIGN POLICY 77 (Winter 1989/90). Copyright © 1989 by the Carnegie Endowment for International Peace.

Terrier antiaircraft missile. Some 1,100 nuclear weapons would be taken off American vessels. These and other short-range, or tactical, nuclear weapons for ocean combat were inventions of the mid-1950s, when the military's infatuation with nuclear weapons spawned everything from nuclear land mines and nuclear bazookas to nuclear-powered rocket engines. The navy, worried about falling behind the air force in the prestigious nuclear field, developed a full spectrum of nuclear armaments for its vessels, including the ASROC, the SUBROC, and the Terrier, as well as nuclear torpedoes, nuclear artillery, and nuclear bombs for delivery by naval aircraft.

Unlike submarine-launched ballistic missiles (SLBMs), which are strategic weapons designed to attack distant land targets, tactical nuclear weapons are primarily meant for attacking other naval vessels and aircraft at short and intermediate ranges. Low-flying, long-range SLCMs [submarine-launched cruise missile] can fulfill both tactical and strategic roles, placing them in a uniquely ambiguous category.

Tactical Weapons Ignored

Once lodged in the navy's arsenal, tactical nuclear weapons rarely entered the public debate over nuclear arms. Since they were based at sea, they aroused little opposition from the arms control community, which was traditionally preoccupied with the more visible nuclear arms race in Europe and parts of Asia. While arms controllers in the United States and Europe pushed for strategic weapons limits in the Strategic Arms Limitation Talks (SALT) I and II and in the Strategic Arms Reduction Talks (START), for limits on intermediate-range land-based weapons in the INF [intermediate-range nuclear forces] treaty, and for a nuclear test ban in a variety of forums, tactical nuclear weapons were largely ignored. By 1988 the U.S. Navy had stocked some 3,660 tactical nuclear weapons for ocean combat.

Ironically, these weapons raised more problems for

the navy than they did for arms control advocates. Exploding a tactical nuclear weapon underwater or in the air would disable the sonars and radars needed to conduct a naval war. Unlike conventional munitions, they require cumbersome security and inspection procedures and take up scarce stowage space that could otherwise house conventional weapons, which, unlike nuclear weapons, can be used with relative impunity in Third World conflicts. As conventional weapons grew more sophisticated and precise, stocking ships with tactical nuclear weapons seemed to make less and less sense.

By the late 1980s these weapons were causing the navy a new set of headaches. As they grew older, they became more expensive to maintain. Having them aboard almost all U.S. combat vessels made them an increasingly popular target for antinuclear protestors. And throughout the Reagan years, Congress refused to fund nuclear replacements for the ASROC, SUBROC, and Terrier systems, largely because of skepticism about their usefulness in combat and in deterring attacks on U.S. ships.

Finally, the navy acknowledged a point that some strategists had long stressed: Tactical nuclear weapons are more useful to the Soviet navy than to U.S. forces because the Soviets could compensate for their inferior conventional capabilities by scattering the oceans with mushroom clouds. "There is a recognition that if there is a nuclear war at sea, we have got more to lose than the Russians," said Vice Admiral Henry Mustin in April 1989, several months after retiring as deputy chief of naval operations for plans, policy and operations. "The concept of a nuclear war at sea is a concept whose time has passed."

The decision to retire these weapons unilaterally was a remarkable admission by the navy that these types of nuclear weapons were deeply problematic for ocean combat. But the fact that the navy purposefully kept this decision from the public—and made no attempt to bargain for Soviet reciprocation—points to the navy's paradoxical aversion to arms control.

"Tactical nuclear weapons are more useful to the Soviet navy than to U.S. forces."

In December 1985 Gorbachev retired Fleet Admiral Sergei Gorshkov, who had commanded the Soviet navy for 30 years. Faced with the need to cut the navy's budget drastically, Gorbachev needed a naval leader ready to place his service at the disposal of a new national defense policy built upon smaller military budgets, greater cooperation with the West, and fewer military commitments abroad. The removal of Gorshkov coincided with the opening of an assertive Soviet campaign to build support for naval

arms control. In a series of speeches and press interviews, Gorbachev and his top advisers have offered a plethora of proposals for bilateral and multilateral naval cutbacks, including

• withdrawing U.S. and Soviet forces from the Baltic Sea, the Mediterranean Sea, and the Indian Ocean;

• limiting the navigation of nuclear-armed ships so that "the coast of any side" would not be in range of nuclear weapons;

• banning naval activity in international straits and major shipping lanes;

• dismantling both the Soviet naval facility at Cam Ranh Bay in Vietnam and the U.S. Navy base at Subic Bay in the Philippines;

• decommissioning 100 Soviet submarines in exchange for the removal of five to seven U.S. aircraft carriers from service; and

• establishing sanctuaries for ballistic missile submarines in the Baltic, North, Norwegian, and Greenland seas and the Pacific and Indian oceans, where they could patrol without being hunted.

Resolution of Naval Issues

The Soviets also contend that a START treaty or a treaty on reducing conventional forces in Europe should be linked to the resolution of certain naval issues. Soviet START negotiators argue that limiting long-range ballistic missiles would be pointless unless curbs are also set on long-range nuclear-armed SLCMs. Although SLCMs are much slower than ballistic missiles, their range, accuracy, and stealth enable them to attack many of the same targets.

In the Conventional Forces in Europe (CFE) negotiations, Soviet diplomats have agreed to deep, asymmetrical cuts in land-based forces, where the Warsaw Pact enjoys a numerical advantage. They argue, however, that the West should reciprocate by negotiating cuts in naval forces, where the West is ahead. "Sooner or later these negotiations [on naval arms] will have to be conducted," Marshal Sergei Akhromeyev, chief military adviser to Gorbachev, told the U.S. House Armed Services Committee in July 1989. "Otherwise, the entire arms control process will break."

Coupled with these Soviet overtures and warnings has been a drastic reduction in Soviet naval activity. Between 1984 and 1987 the Soviet navy reduced its surface ship deployments out of coastal waters by one-quarter and its submarine deployments by almost half. Large naval exercises far from Soviet shores dropped off dramatically. The size of the Soviet fleet has also begun to diminish, as scores of older vessels are retired. The Soviets removed more ships from active service in 1988 than in any other year in recent history. The ship-junking effort has even taken on a *glasnost*-era flair: In May 1989 the Pepsi-Cola company agreed to take a cruiser, a destroyer, a frigate, and 17 submarines as scrap in payment for its products sold in the USSR. . . .

While Soviet naval reductions have been dismissed at the Pentagon as a passing nuisance, Soviet naval arms control proposals have been treated as a direct attack. According to Carlisle Trost, the chief of naval operations, "Despite the seeming sincerity, love of peace, and desire for friendship radiating from these Soviet initiatives for naval arms control, the real motive is to reduce an area of disadvantage at little cost to themselves. . . . If the Soviets accomplish even one of the goals of their present campaign [for naval arms control], our diplomacy will have suffered disaster." Even confidence- and security-building measures (CSBMs), such as notifying the other side in advance of large naval exercises and exchanging observers, are seen by the U.S. Navy as detrimental. In April 1989, the deputy chief of naval operations, Vice Admiral Charles Larson, told Congress that CSBMs at sea would be unacceptably intrusive, set a bad precedent, impinge on the doctrine of freedom of the seas, inhibit the navy's missions outside the European theater, possibly violate international law, and "weaken the West's deterrent posture and consequently decrease Western security." Not surprisingly, the U.S. Navy declined a Soviet offer in June 1989 to observe the USSR's naval exercises for the first time. The exchange of observers for land-based exercises, meanwhile, has become common.

The navy's opposition to arms control is anchored both in America's history as a maritime nation with overseas interests that could only be maintained with naval power, and in the navy's status as an elite, independent service that shuns external constraints. The navy believes that any form of arms control would almost certainly clip its hard-earned advantages over the Soviet Union. Moreover, the navy argues this advantage is justified by America's geographical dependence on the oceans for both national security and international trade. This perception of the United States as a nation reliant on sea power makes any restrictions on naval arms hard for the Pentagon to swallow. . . .

The Standoffish Navy

The Reagan and Bush administrations have thus far backed up the navy's opposition to joining in arms control, with one significant exception. In December 1987 the Reagan administration sidestepped navy objections and accepted a Soviet demand to discuss limits on SLCMs in START. Since then, SLCMs have become one of the main sticking points in the negotiations.

Since it is difficult to distinguish between SLCMs armed with nuclear warheads and those armed with conventional warheads, the first Soviet proposals called for a cap on both variants. In July 1989, however, the Soviet Academy of Sciences and the New York-based Natural Resources Defense Council jointly tested passive radiation detectors that could pick out missiles with nuclear warheads. Following these well-publicized experiments, the Soviets proposed a ban on all nuclear SLCMs.

"The Soviets removed more ships from active service in 1988 than in any other year in recent history."

After agreeing to consider SLCMs in START, the United States has agreed to little else on this issue. Its standing position is that each side should announce how many SLCMs it has deployed, then proceed without further restrictions. American negotiators have backed the position of the U.S. Navy that limits on nuclear SLCMs would be impossible to verify.

During his September 1989 visit to the United States, Soviet Foreign Minister Eduard Shevardnadze indicated that the USSR would not allow the dispute over SLCMs to hold up a START treaty. In a joint statement with Secretary of State James Baker, Shevardnadze said that "these weapons could be limited outside of the text of a START treaty." He also suggested that SLCM limits be addressed "in a broader arms context" and appealed to the United States to help resolve the problem of verification. Baker, however, reiterated U.S. doubts that SLCM restrictions could be verified, and "noted its long-standing view that there are serious problems involved in any discussion of the limitation of naval arms."

The navy's opposition to measures that would diminish its edge over the Soviet navy is understandable. But its aversion to arms control overlooks some critical distinctions. There are four types of naval arms control available to the superpower navies. The first involves geographical constraints, such as bilateral limits on naval forces or naval operations in certain regions. The second type entails numerical limits on the vessels of each navy. The third addresses tactical nuclear weapons. The fourth includes CSBMs that would make the activities of each navy more transparent to the other.

The Soviet initiatives for naval arms control touch on all four areas; some are of little potential interest to the United States. Among the least promising offers are those that suggest geographical constraints, such as a withdrawal of warships from the Baltic Sea, the Mediterranean Sea, and the Indian Ocean. While these actions might make the world a safer place, any restriction on the movement of naval forces is anathema to the U.S. Navy's traditions and philosophy and thus has little chance of meeting with American approval. Another form of geographical naval arms control is the Soviet proposal to establish sanctuaries for ballistic missile submarines in which they could patrol without being hunted by planes, ships, or other submarines. Since this plan would nullify the U.S. lead in antisubmarine capabilities, it understandably garners little enthusiasm at the Pentagon. A third Soviet proposal for geographical arms control, the offer to give up the relatively small Soviet naval facility at Cam Ranh Bay if the United States abandons its large base at Subic Bay, is too lopsided to be taken seriously and was probably produced for propaganda value.

"The navy believes that any form of arms control would almost certainly clip its hard-earned advantages over the Soviet Union."

The second area of naval arms control, limiting the number of vessels in each navy, may be useful in the future but seems less than urgent at a time when both navies are shrinking fast anyway. The Soviet proposal to decommission 100 submarines in exchange for the removal of five to seven U.S. aircraft carriers is partially being realized through budget cuts on both sides. The Soviets have pushed to include naval forces in the CFE talks in hopes of offsetting their large proposed cuts in land-based forces with parallel cuts in NATO naval vessels. If previous Soviet negotiating behavior is any guide, the Soviets will eventually drop this demand, perhaps in exchange for other concessions.

Reducing Tensions

Yet in other areas Soviet offers merit serious consideration. Simple CSBMs would reduce tensions at sea as they have on land. Banning nuclear-armed SLCMs might be of greater benefit to the United States than to the Soviet Union. Short-range nuclear SLCMs are the major striking force of the Soviet navy. Eliminating them would help ensure the survival of the U.S. Navy in a nuclear confrontation. And while the United States at present holds a technological lead over the USSR in long-range SLCMs, when the Soviets catch up it could prove disastrous for the United States, which has a much higher concentration of its population and industrial capacity close to shore where it is vulnerable to attack from SLCMS. The advantages of banning nuclear SLCMs were articulated in the Discussion Group On Strategic Policy, whose members include Senators Sam Nunn (D-Georgia) and John Warner (R-Virginia) of the Senate Armed Services Committee and Representative Les Aspin (D-Wisconsin) of the House Armed Services Committee, in their January 1989 report, "Deterring Through the Turn of the Century": "In naval warfare, nuclear SLCMs would allow the Soviets to compensate for poorer accuracy. The U.S. advantage in accuracy should enable us to succeed—indeed, to prevail—in a conventionally armed SLCM contest."

An increasing number of naval experts— most prominently INF negotiator Paul Nitze and former CIA [Central Intelligence Agency] Deputy Director Admiral Bobby Inman—have gone further and called for a ban on all tactical naval nuclear weapons, including SLCMs. Such a move could help protect the superiority of the U.S. Navy, freeing it from the danger of nuclear attack. It could also end the danger of nuclear war at sea, address Soviet concern about SLCMs in START, and improve relations between the United States and a critical group of NATO and Pacific allies facing strong anti-nuclear pressures, including Iceland, Norway, Denmark, Spain, Greece, Japan, the Philippines, and, of course, New Zealand.

The U.S. Navy has blocked any form of arms control at sea out of concern that it will level off America's considerable naval advantages over the Soviet Union and tie America's hands in the Third World. But in rejecting measures that might disproportionately benefit the Soviet Union or run counter to naval traditions, it has also rejected measures that would bring equal, if not greater, benefits to the United States. The Soviet leadership may remain flexible and eager for naval arms control for years to come. If it does not, however, the United States may be losing an opportunity to improve stability on the high seas and eliminate the troublesome folly of tactical naval nuclear weapons.

Amid the superpower rapprochement of the late 1980s, the U.S. and Soviet navies have both seen their public esteem and political fortunes decline. A pair of Soviet submarine disasters off the coast of Norway in 1989 left the public wondering whether the Soviet navy is a greater danger to its own sailors than to the freedom of the West. For the United States, a trio of disasters in the Persian Gulf in 1987 and 1988—the missile attack on the frigate *Stark* by a confused Iraqi pilot, the crippling of the frigate *Samuel B. Roberts* by a World War II-type mine, and the downing of an Iranian passenger plane by the cruiser *Vincennes* —tarnished the image of competence the navy tries hard to project. Both the fatal explosion aboard the World War II-era battleship *Iowa* in April 1989 and the navy's controversial investigation of the incident further sullied its reputation.

Similar Strategies

Their fates oddly linked, the two navies have adopted similar strategies to cope with shrinking budgets. Both have retired older ships earlier than planned, while keeping the acquisition of new vessels largely on schedule. Both navies have also made important cuts in their overseas deployments and in the size of their naval exercises.

The U.S. Navy has fought a public relations battle to discount Soviet naval cutbacks and to elude all arms control and CSBM discussions. The Soviets, meanwhile, have used their less capable navy to highlight an area of American superiority and to press the United States with a fusillade of public naval arms control proposals. Thus the Soviet government has made a political virtue of economic necessity. By contrast, the Bush and Reagan administrations, hampered by their navy's abhorrence for arms control, have taken pains to play down the ongoing cuts in U.S. naval strength and operations for fear of catching the attention of arms control advocates.

"The Soviet initiatives for naval arms control . . . are of little potential interest to the United States."

Ironically, the Soviet Union has lagged behind the United States in recognizing the obsolescence of tactical nuclear weapons. By retiring three aging nuclear weapons systems, the U.S. Navy is unilaterally cutting its tactical nuclear arsenal by almost one-third. Yet tactical naval weapons are the only class of nuclear weapons not subject to any current or foreseen arms control negotiations. This has led to an odd stalemate. The U.S. Navy has little faith in tactical nuclear weapons but even less faith in negotiated arms cuts; the Soviet government is anxious for almost any type of naval arms control but has said relatively little about tactical naval nuclear weapons. The Soviet proposal to ban nuclear SLCMs may be a harbinger of change in Soviet thinking.

Ebbing of the Cold War

Despite the American aversion to naval arms control, the naval arms race has slowed, albeit unevenly. Even if the U.S. Navy can avoid sitting across a negotiating table from the Soviets, it cannot avoid the ebbing of the Cold War and the concomitant shift in resources away from the military. The navy's goal of 15 aircraft carrier battle groups has been abandoned; 13 or 14 active carriers now seem likely. The success of New Zealand in banning visits by nuclear-armed warships has encouraged antinuclear forces in Europe and the Pacific, who are now determined to make the visits of nuclear-capable ships a major domestic political issue in more than a dozen countries.

Even the once-bold Maritime Strategy today has been diminished to a vague blueprint for addressing a variety of potential crises around the world. Shorn of its major opponent, the U.S. Navy seems unable to adjust to an era of improved superpower relations. Once again, the navy is searching for a mission.

The growing political and economic constraints on the two navies have forced a kind of de facto naval arms control. But this is hardly preferable to negotiated naval arms reductions. It has allowed the navies to cut where they are weakest and to continue building weapons and vessels that the other side considers threatening. It is time that the arms control process be extended to the superpower navies before an opportunity is lost to wind down the risks of the Cold War at sea.

Michael L. Ross is a former congressional aide for foreign and defense policy. He is now the international communications coordinator for the Nuclear Free Seas campaign run by the international environmental group, Greenpeace.

Negotiating Cuts in Naval Weapons Would Harm U.S. Interests

Carlisle A.H. Trost

Editor's note: The following viewpoint is a speech given by U.S. Chief of Naval Operations, Carlisle A.H. Trost at the Leningrad Naval School in the Soviet Union.

In May of 1972, our governments signed the agreement on the prevention of incidents on and over the high seas between our navies, an agreement that has served both navies well. In like manner, my visit to your country is part of a much broader program to enhance mutual understanding and reduce tensions between our two countries. It is a privilege for me to be the first chief of a U.S. military service to visit in continuation of the program which has seen exchange visits by our respective ministers of defense and senior military leaders.

The thirteenth of October 1989, will be an equally special day for me and the U.S. Navy. When I assumed the position as Chief of Naval Operations on the steps of our own Naval Academy, I certainly did not think that I would be celebrating the 214th birthday of the United States Navy here in the Soviet Union, visiting ships in your fleet, touring aircraft, and meeting officers and men in your Navy. In itself I think that my being here marks a milestone in the progress that our two countries have made in just the past few years toward improving bilateral relations and easing some of the tension that has characterized our dialogue for four decades. The exchange of top military officers, port visits by ships of our fleets, and cultural exchanges all serve to promote a better understanding between us. We should take full advantage of every opportunity to learn more about one another so that we can foster resolution of differences, exploit similarities, and develop trust.

As my Navy celebrates its birthday tomorrow, I think it is interesting to note that both of our modern navies had very humble beginnings. In each case, our

Speech delivered in the Soviet Union at the Leningrad Naval School on October 12, 1989.

navies began with small boats crewed by a handful of daring and dedicated sailors, who were tasked to sail in harm's way against a much more powerful adversary. The early fleet of Peter the First, with victories on the nearby lakes, Ladoga and Peipus, and the River Neva, established a heritage of excellence that remains apparent in your Navy today.

Humble Beginnings

In my country's struggle for independence 70 years later, an able young naval officer named John Paul Jones would carve a similarly glorious tradition of victory at sea, also against a vastly superior foe, and become the father of the United States Navy. Today he rests in a crypt beneath the chapel dome at the U.S. Naval Academy in Annapolis surrounded by mementos that record his accomplishments. Among them is a Ruby Red Cross, the Order of St. Anne, which was awarded to him by Catherine the Second, for his service on the Black Sea as a Rear Admiral in the Russian Navy. I should think that few men, if any, could achieve such distinction and rank in two navies and serve with such valor. You and I as military men can appreciate such things.

Being military men, inheritors of these similar beginnings, we share other things in common. In both our countries the military is subordinate to civilian leadership. And while we are properly excluded from making national policy, we must execute that policy within the political framework that our governments prescribe. It is incumbent upon us to be fluent in global dialogue. This is particularly germane in the case of naval officers because our business is at sea in the international arena day in and day out. Toward this end, military colleges, universities, and academies that train our officers are invaluable national resources. The time we spend ashore should be used with the same dedication as the time we spend at sea.

This morning I want to discuss several topics so

that you know exactly how I feel on some subjects that I think are of our mutual interest. My comments are of value to you only if I speak with complete candor, naval officer to naval officer. Empty rhetoric or mindless propaganda benefit no one. If we are to succeed in our efforts to reduce tension, we must each understand the position the other takes on issues of mutual interest or concern.

Era of Change

We are living in an era of enormous change. . . . However. . . there are three things that I think will remain constant for the foreseeable future. First, the United States is a nation that relies on the sea for its economic and political livelihood. Second, the Soviet Union is the only nation in the world that has the capability not only to challenge our way of life, but perhaps even to destroy its very existence. And third, independent of the actions of the United States and the Soviet Union to reduce tensions, the rest of the world is becoming more economically inter-dependent, while concurrently becoming more independent politically and militarily. For this reason, I think we can expect to see a relative decline in the influence that the Soviet Union and United States exert on the actions of individual nations. With those thoughts in mind I am going to speak about the U.S. maritime strategy, naval arms control, and the future international security environment as I see it. First U.S. maritime strategy.

We published our maritime strategy in open literature. Since then this document has been the subject of controversy both in my country and around the world. The U.S. maritime strategy is the maritime component of the overall U.S. national security strategy. It is not a war plan. Nor is it a document that outlines a predisposition of naval forces to wage war. The maritime strategy is a concept, repeat concept, of operations for the effective global employment of naval forces to protect the interests of the United States and our allies and support our national policy objectives. It is the same strategy that the United States has pursued in the name of peace for the past forty years, and is based on three fundamental tenets.

"The purpose of any negotiation for arms control must be a meaningful improvement in the security posture for all of the participants."

The first tenet is deterrence. Its purpose is to deter any potential adversary from either attacking the United States and our allies, or attempting to undermine the economic and political interests on which we rely. The strategy is sufficiently broad to cover the employment of naval forces across the entire spectrum of conflict, ranging from global nuclear or conventional war down through regional conflicts in peacetime and crisis.

Secondly, the strategy is built around a network of alliances. Since World War Two the United States has established agreements with over forty countries to provide mutual security for common defense. The strength is not in the military power of any one individual but the combined strength of the alliance in which each member shares the burden of defense. Granted, the United States is the leader in these alliances. In the coming years I expect to see many of our allies begin to assume greater responsibility for the common defense. This may be particularly true with NATO [North Atlantic Treaty Organization]. I think it is important to note that in the forty years that NATO has stood united and kept the peace in Europe, there has not been a single aggressive act by any one of its members against a nation in the Warsaw treaty organization.

Third, and probably least understood and possibly most worrisome to potential adversaries, is the premise of forward deployment.

Reliance on Overseas Trade

Now, some argue that forward deployment poses an offensive threat. Among them is Marshall Akhromeyev, who, on a visit to the United States in the summer of 1988, looked me in the eye and said, "You, you're the problem. Your Navy and bases surround my country and threaten the security of the Soviet Union." My response then and now is the same. The United States strategy is not intended to threaten anyone. Geographic reality is such that many of our allies and trading partners are located on the periphery of the Eurasian landmass. If the United States is to effectively participate in mutual defense of our own and our allies' interests, it is imperative that we have forces deployed close to regions of potential conflict. In the last several years the United States has placed increased emphasis on the role of naval forces in forward deployment because of the changing international environment. Since 1950, there has been a 60 percent decrease in both the number of overseas basing facilities and number of host countries for our forces. But, there has been no decrease in our overseas interests. Quite the contrary, the United States relies more heavily on overseas trade than ever. Forward deployed naval forces give us the flexibility and mobility to continue to protect these interests. They are only a threat to someone who would intend to threaten our interests or those of our allies.

The second topic I want to address is naval arms control. The purpose of any negotiation for arms control must be a meaningful improvement in the security posture for all of the participants. While force reductions may produce reduced governmental spending as a by-product, that cannot and should not be the principal focus. The goal must be improved stability. Unfortunately, I think our respective definitions of stability are somewhat different. The writings, speeches, and proposals of some of your leaders lead me to believe that you view stability as being synonymous with predictability. Predictability, if it means that restrictions are placed on the movement and composition of ships on the high seas, can foster a climate ripe for deceit and adventurism. My definition of stability focuses less on attempting to limit, through agreement, a potential adversaries' options, and more on the well understood, historically demonstrated national policies of the countries involved and the deterrent effect of these countries operating viable forces in their regions of interest. The principles that may govern stability on land cannot be translated to apply equally on the high seas. Navies don't occupy territories. All nations have free and equal access to the seas. Naval forces by virtue of their mobility and global access can be concentrated to deter and then just as quickly depart without the adverse implications or difficulties involved in the use of land forces. And while naval forces don't singularly win wars, their absence can certainly result in the loss of wars, especially if one nation is dependent on the sea. On that point I'm sure we are in agreement.

Island Nation

Many of your leaders have stated that the single most significant obstacle to the continuing improvement in relations between the U.S. and the Soviet Union is our reluctance to entertain the inclusion of naval forces in overall arms control talks. That may be so from your perspective, but in my view such statements fail to recognize the fundamental differences between our respective geographies and national security requirements. The United States is an island nation. Two of our states, Alaska and Hawaii, are separated from the mainland of the United States by thousands of miles of ocean. The vast majority of our trade is with nations across the great expanse of the Atlantic and Pacific Oceans. And again we are critically dependent on this trade for economic survival.

Contrast that picture with your own country. You are virtually self sufficient in basic energy and strategic requirements. The states of the Soviet Union are all on the same landmass. Your principal allies and trading partners are also on the same landmass. Seaborne trade for the Soviet Union is not a matter of national survival.

So, when viewed from this balanced perspective, I strongly feel that my country's reluctance to enter into naval arms reductions is justified by the facts and is a prudent and rational position.

Let's look at a few specific proposals that members of the Warsaw treaty organization and some others have offered. One calls for the exclusion of anti-submarine capable forces from specific security zones. Another calls for the exclusion of all naval force

activity in certain strategic straits and high density shipping lanes. Other proposals seek to limit the scope and number of naval exercises, and when such exercises occur, provide for advance notification and the embarkation of observers. Still others seek to restrict the movement of ships that may be armed with nuclear capable weapon systems. In each case, I interpret these proposals as attempts to abrogate commonly accepted international law with respect to freedom of the high seas. Any one of these would result in the inability of my Navy to protect the global interests of the United States or to deter aggression. Naval forces must be free to operate when and where deterrent presence is required, and operate unimpeded by restrictive sanctions.

"I strongly oppose any negotiation that would impose undue restrictions on cruise missiles at sea."

To those who would argue that my position on these measures is intractable, let me remind them that our two countries already have formal and informal measures which have proven effective in reducing the probability of conflict on the high seas. The Incidents at Sea Agreement of 1972 has enjoyed remarkable success in preventing inadvertent mishaps between U.S. and Soviet fleet units. The Stockholm Accord of 1986 already carries stipulations that require advance notification of naval exercises within specific limits. The Madrid Mandate will expand on the Stockholm agreement to include other navy activities, if such activities are functionally linked with operations on land. When Admiral William J. Crowe visited the Soviet Union, he and General Moiseyev signed an agreement to reduce dangerous military incidents in regions where the armed forces of our two countries routinely operate. These are all sound agreements that result in an increased measure of stability, but do not impinge on any nation's free use of the high seas.

Sea Launched Cruise Missiles

Another topic that seems to surface frequently when arms control is mentioned is sea launched cruise missiles. I understand that the Soviet Union views the U.S. sea launched cruise missile capability with concern. You, as military men and learned strategists, can appreciate it when I say that it is intended to concern you.

More than twenty years ago your Navy embarked on a weapons building program whose sole purpose was to target and counter U.S. aircraft carriers. The Soviet Navy developed a powerful naval air arm, potent submarine force, and blue water surface force, all capable of carrying large numbers of cruise missiles, many with nuclear warheads—and each one targeted against our aircraft carriers.

In response, we felt we were left with no option but to develop a capability to disperse the surface and land strike assets that were previously concentrated only in our manned aircraft aboard carriers. Hence, the sea launched cruise missile was developed and is now deployed on surface combatants and submarines. In addition to complicating an adversary's targeting effort, the cruise missile gives fleet and battle group commanders another asset for a measured response, and one which does not endanger airmen's lives in striking targets at sea or ashore.

"The submarines your leaders propose to retire . . . have surpassed their useful service life and will be retired anyway."

I strongly oppose any negotiation that would impose undue restrictions on cruise missiles at sea. Contrary to what some may say, I believe that compliance with restrictions would be unverifiable without unacceptably intrusive inspections. I noted with interest some articles that appeared in August 1989 concerning the verification experiment conducted on board one of your slava cruisers. I'm referring to the experiment that was jointly sponsored by the Soviet Academy of Sciences and the Natural Resources Defense Council, the latter being a group of scientists and academicians who are not official representatives of the United States, but nonetheless technically knowledgeable. In essence, they concluded just what I said, that unintrusive verification is impossible using the tested techniques.

But more importantly, from my perspective, limits or reductions on cruise missiles would again make the U.S. Navy's seaborne strike capability reside solely in our aircraft carriers. And, then, your cruise missiles would again be aimed primarily at our carriers. This poses unacceptable risks to our ships, our people, and would severely inhibit my Navy's ability to protect our global interests.

Aircraft Carriers

That brings me to the last topic on arms control I intend to discuss—U.S. aircraft carriers. Some of your country's leaders have suggested that the United States should retire or place in storage half of our aircraft carriers in return for your retirement of about a hundred of your submarines. Such proposals do not reflect an understanding of the basic differences in economic and political dependencies between our countries. The aircraft carrier is the backbone of the United States Navy. When combined with supporting surface combatants and logistics ships, it provides a mobile, flexible, and self-sufficient base to protect our

interests and deter would-be aggressors. We have seen examples where the presence of a carrier battle group positively influenced an otherwise potentially volatile situation. The *USS Nimitz* steamed off the coast of Korea during the Olympiad in Seoul. Prior to the games there had been much rhetoric from the North Koreans about interrupting the games with violence. The presence of the *Nimitz* strongly discouraged the North Koreans from following rhetoric with action. Carrier battle groups on station in the North Arabian Sea and Indian Ocean have added a strong measure of deterrence to keep the fragile cease-fire between Iran and Iraq intact. The presence of the *USS Coral Sea* and American battle groups in the Eastern Mediterranean halted the barbarous threats to murder more U.S. and foreign hostages being held captive by state sponsored terrorists in the Middle East.

The United States currently maintains fourteen deployable aircraft carriers. At fourteen, we are barely capable of maintaining our peacetime commitments in regions of the world where the stability they bring is required for peace. I must emphasize that every nation in the world community, not just the United States, benefits from the forward deployed presence and resulting deterrent effect of our carrier forces.

And let me add that fourteen carriers are all front-line operating units, unlike the submarines your leaders propose to retire in exchange. It is clear that most of these submarines have surpassed their useful service life and will be retired anyway.

Future Security

The last topic area I want to briefly address is the future security environment as I see it, and the implications it has for naval forces. As I said earlier, the world order is changing and many of the changes we see today may continue independent of actions by the United States or Soviet Union. We are witnessing a dispersion of power centers with a greater emphasis on economic influence than on military power. The relative world position of the superpowers is decreasing. The improving relations between our two countries may result in an overall reduction in the amount of money that both of us spend on our militaries.

But there are some other things going on in the world which are not so positive. The proliferation of sophisticated weapon systems among many nations in the world should trouble everyone. We've seen the indiscriminate use of the chemical weapons by Iran and Iraq. Many other countries are building facilities to manufacture their own chemical weapons or are trying to buy them from others. Many nations that can't feed their hungry populations are buying or building cruise missiles. By the year 2000, some intelligence estimates predict that 15 or more countries will have the capability to produce and launch ballistic missiles. The prospects for the proliferation of nuclear weapons are not much better.

Terrorist groups, many of which are sponsored and supplied by legitimate states, continue to be the scourge of the civilized world. International drug cartels are getting rich at the expense of young people all around the world. In both instances, the values of human life and decency are absent.

"We, the military, must remain ready to defend our nations' security."

Before your questions, I'll conclude my remarks with this observation. All of us in this auditorium are military men. More than anyone else we have seen and understand the suffering and pain of war. Our governments' principal charge to us is to deter war so that the warfighting skills we have trained long and hard to master will never be required. I am hopeful that this new era of understanding between our two countries will result in a world where our successors will not feel threatened by any nation. But the world has a long way to go to meet that goal. Trust and confidence in the intentions of other nations, including reluctance to use force to attain national goals, comes with time and corresponding actions that reflect those qualities. Our elected governments will control the speed of these developments in our two countries, and hopefully they can influence the rest of the world community to direct their energies in the same direction. In the interim, we, the military, must remain ready to defend our nations' security interests. We must do so not by fancifully trying to assess the intentions of a potential challenge or threat, but by assessing the reality and true capability of those who may pose a threat. The citizens of our countries deserve nothing less.

Carlisle A.H. Trost is chief of naval operations of the United States Navy.

"The aim of arms control is to increase international stability and thus improve the security of every state."

Arms Control Will Lead to Peace

Wolfgang Altenburg

The concept of arms control implies agreements between states to regulate their military potential. In practice, this may mean restrictions on manpower levels, deployments, states of readiness, or types of military forces, weapons or facilities.

The aim of arms control is to increase international stability and thus improve the security of every state.

Throughout the history of warfare, surprise attack has remained a very important principle of war which has often conferred enormous advantage. Such an attack could not only severely damage an opponent but might also neutralize or destroy much of his capacity to strike back. In short, a surprise attack might not only win the war but also lessen the casualties suffered by the attacker. Put in boxing terms, the first blow may be the knockout punch. Obviously, forces which are structured for offensive action are in a much better position to launch a surprise attack than those of a purely defensive nature. As things currently stand, the Warsaw Pact remains structured for surprise and offensive action whilst NATO [North Atlantic Treaty Organization] is not. Things in the Soviet Union may be changing in this respect but the basic truth of this statement remains valid.

The very act of striking first gives great advantage, and recognition of this can indeed be an inducement to pre-emptive action. Take the example of two gun-fighters facing one another. Each knows that he *might* be able to kill his opponent but there is no certainty that he would do so. By shooting first, one of them would have the advantage—provided, of course, that he is accurate! Indeed, the advantage of shooting first might encourage this very action. Equally, pre-emptive action might be seen as self-defence if one felt sufficiently threatened and at risk. Such was the argument often legally accepted by courts of law in the American West! Using the analogy in arms control, our aim must be to reduce the chances of one gun-fighter trying his luck, and to produce a situation in which there would be no advantage in 'jumping the gun' and no expectation that a potential enemy would do so either.

Conventional Forces

I have already stressed that the central theme of arms control is to increase stability—indeed arms control is invalid unless it does so. In the NATO context, it is natural to feel threatened by a neighbour who is numerically stronger and seemingly poised for what could be a surprise attack. Looking at conventional (non-nuclear) forces, the Warsaw Pact is much stronger than NATO and these forces still appear to be deployed for a surprise move against the Alliance. Changing this situation is the highest priority for our negotiations on Conventional Armed Forces in Europe (CFE). In reality, arms control depends on self-interest. To be effective, it must be attractive for all participants to establish and maintain it. Instruments which establish arms control agreements also require consensus on two vital principles. Firstly, there must be agreement as to the exact *limitations* imposed. Secondly, there must also be agreement on the way in which *verification*—which is absolutely vital—is to be carried out.

Fundamental to negotiating an arms control/reduction agreement is thus the establishment of an agreed data base. If parties striving to negotiate an arms control treaty cannot agree on basic data then the necessary building blocks for agreement are absent.

Now let us look at the matter of verification. Although it may no longer be necessary for a 'Mata Hari' figure to elicit secret details about the enemy given our modern national technical means, such details still require to be checked 'on site'. Effective inspection is more vital to Western states, who must

Wolfgang Altenburg, "Arms Control and the Future," *NATO Review*, August 1989. Reprinted with permission.

penetrate closed societies, than those in the East, who have easy access to Western press, radio, TV, parliamentary/congressional reports and even public statements. In short, the value and costs of inspection are closely linked to the open or closed nature of the target society. Nobody should be under the illusion that verification is a cheap option. To send inspection teams to check compliance is a very time-consuming and expensive undertaking. Effective arms control is neither easy nor cheap.

And it is not 'foolproof'! If one hundred per cent certainty is a prerequisite of verification then arms control would just not be feasible. We all know that no system depending on human judgement is infallible. But in practice, one hundred per cent certainty of verification is not required. It may be enough that violations are likely to be discovered, but the odds of discovery should be high enough to ensure that the risks of violation are simply not worth it. Reconnaissance satellites, on-site inspections and other measures must be designed to maximize the chances of this.

International Moves

NATO has consistently sought greater security through arms control measures. In 1967, the Alliance formally accepted the Harmel Report which described NATO's two main functions. Firstly, the Alliance has '. . . to maintain adequate military strength and political solidarity to deter aggression and other forms of pressure, and to defend the territory of member countries if aggression should occur'. Secondly, it must '. . . pursue the search for progress towards a more stable relationship (between East and West) in which the underlying political issues can be solved.' Arms control is a vital part of NATO's approach to international security. But until recently, the Soviet desire to reach arms control agreements was obscure to say the least. Why then has the Soviet approach changed?

Certainly, I believe we can take Mikhail Gorbachev seriously. I have no doubts that he is a brave and inspiring man who is genuine in his wish to alter Soviet relationships with the West. Paramount among these relationships is that of economics. The Soviet economy is apparently faltering so badly—especially when compared to the West—that there is perhaps one of the greatest crises ever facing the Soviet Union. It is not NATO's military hardware that Mr. Gorbachev and his colleagues currently fear most, but rather the very poor standard of living in the Soviet Union vis-a-vis that of the West. I read in *Time* magazine that a Soviet farmer can feed between seven and nine people on an area of land from which a Dutch farmer could extract food for 112! That gives an example of what *perestroika* and *glasnost* need to remedy.

Mr. Gorbachev must deliver on the economy and, as part of that delivery, less resources must be devoted to the Soviet defence budget. Even if Mr. Gorbachev does reduce his defence budget, an intention he has publicly announced, it seems that the Soviets will still be spending about 14 per cent of GNP [gross national product] on the military. By comparison, NATO defence budgets are all less than half of that, with the highest being under seven per cent and the lowest just over one per cent of GNP. But at least the Soviets are now moving in our direction and we should encourage this as much as possible. Nonetheless, we should also recognise the opportunity it gives us to obtain binding and verifiable arms control agreements which contribute to increasing stability throughout Europe and the world.

Aims and Problems

The Comprehensive Concept of Arms Control agreed upon at NATO's May 1989 Summit clearly emphasized the Alliance's aims. At the strategic level, the Allies continue to support the United States' objectives in Strategic Arms Reduction Talks (START)—a 50 per cent cut in strategic nuclear missiles. At the CFE negotiations in Vienna, the Alliance abides by three guiding principles: the achievement of secure and stable balances of conventional forces at lower levels, the elimination of disparities prejudicial to security and stability, and the elimination, as a high priority, of the capability for launching surprise attacks and for initiating large-scale offensive action. The main point of President George Bush's proposal to add aircraft and combat helicopters to the talks, as well as to introduce reductions and ceilings on US and Soviet forces stationed outside their national borders in the Atlantic to the Urals region (275,000 on each side)—now officially a NATO proposal as well—was to prove just how serious the West is on arms control. Despite the international euphoria about changes in the Soviet Union and some troop reductions, I still see at least 28 Soviet divisions deployed in Eastern Europe. By 1991, this may well have been reduced by about six divisions, but even so, 22 Soviet divisions will remain there. By contrast, the United States only has four-and-three-quarter divisions in Western Europe.

"Effective arms control is neither easy nor cheap."

Announcing the possible reduction of United States troops in Europe from 305,000 to 275,000 was thus a signal aimed primarily in two directions: at the Soviet Union and at international opinion. On the one hand, it was intended to force down the Soviet's overwhelming superiority in conventional forces and, on the other, to clearly demonstrate Western resolve and intentions in arms control. It was certainly *not* a signal directed at other NATO members suggesting that they should consider unilateral cuts themselves. To do so, would take the ground from under the feet of our negotiators in Vienna. We must negotiate from strength, not a sinking raft!

Although the vast majority of initiatives and ideas on arms control have come from the West, we frequently get very little credit for them. By contrast, a proposal from Mr. Gorbachev is a 'headline stealer'. . . .

The quicker we reach agreement on conventional force reductions, the sooner might we consider *partial* reductions in short range nuclear forces. NATO has no doubts whatsoever about the value of such weapons within its strategy of flexible response. They remain essential to it but it may be possible for us to consider reducing their numbers—once the threat posed by the overly large imbalance in conventional forces in Europe has been reduced. The key is an early CFE agreement in Vienna.

"Although the vast majority of initiatives and ideas on arms control have come from the West, we frequently get very little credit for them."

I am sure that both sides have similar aims for the CFE negotiations. They each want increased stability at lower levels of military confrontation. But I will mention here two obstacles that need to be overcome on the way to that agreement.

Obstacles to Agreement

To start with, there is the so-called data question. In November 1988, NATO produced its assessment of the conventional force balance in Europe. By January 1989, the Warsaw Pact had replied with its own version. Although there are considerable differences between the two sets of figures, the Warsaw Pact figures are at least the latest example of *glasnost* and are to be welcomed. In the 16 years of the Mutual and Balanced Force Reductions (MBFR) negotiations, the Warsaw Pact consistently refused to produce such data and so, at least, we do have a basis for discussions. To begin with, we need to agree on a common set of working data.

But figures alone will not be enough, even though it was Lenin who once declared, 'quantity has a quality all of its own.' A big question remains over operational capabilities—the effectiveness of the differing formations which make up the forces of both sides. This dilemma not only includes fighting qualities but also strengths, readiness, training, equipment and deployment. There can be no doubt that the power of the Warsaw Pact's conventional forces in Europe is significantly greater than that of

NATO's, and these forces would need to diminish mightily before any substantial Western reductions could be envisaged.

The Intermediate Nuclear Force (INF) Treaty signed in December 1987 may well prove to have been a turning point in the history of arms control. For the first time, asymmetric reductions and stringent verifications were accepted by the Warsaw Pact. Hopefully, the INF agreement will act as a catalyst to further substantive arms control negotiations.

To end, I return to the twin thrusts of the Harmel Report that clearly shows arms control to be an integral part of NATO's security policy. The military offers advice on negotiating positions, always with the aim of maintaining NATO's security. But our attitude towards such negotiations is very much influenced by former Belgian Foreign Minister Pierre Harmel, who stated: 'We must aim to maintain peace and security in East-West relationships, and any lower levels of weapons arsenals should be welcomed—provided it is not destabilizing and increases rather than decreases security.' In arms control, what Harmel said over 20 years ago remains completely valid today.

General Wolfgang Altenburg is the chairman of the Military Committee of the North Atlantic Treaty Organization.

"Only a disarmed world can bring security and survival."

Arms Control Alone Will Not Lead to Peace

John M. Swomley

A new day has dawned in human history, making disarmament a practical political possibility. Only once before in human history, with the Emperor Ashoka in India, has a major military power renounced imperialism, disavowed future aggression and unilaterally started the process of reducing its weapons to defensive status, as the Soviet Union has now done.

Unfortunately, the American peace movement is unprepared to take advantage of this new political opportunity. Although it has never been stronger at the local level, it lacks strategic and ideological leadership at the national level.

Segments of the national anti-war movement have concentrated on opposing development or deployment of specific weapons such as moveable missiles (MX). Others have worked to achieve the acceptance of a specific treaty, such as Strategic Arms Limitation Treaties (SALT I and II), even though the limits set there may have been greater than the current levels. We should beware of largely cosmetic treaties.

In general it is more important to shape the future thinking and politics of a nation with respect to disarmament than it is to seek the abolition or limitation of a specific weapon. However, the dramatic use of nonviolent action by the Berrigans and others on the Catholic left, even when focused on specific weapons, could not be ignored by most right wing bishops or moderate anti-war groups, and thus helped to shape the future. On the other hand, such action is no longer innovative enough to make news, and does not strategically address the problem of disarmament.

A second weakness of the anti-war movement is its emphasis on justice as the way to peace. This was a major mistake of the early communist movement. It asserted that the way to end war was to end

capitalism. The pacifist analysis stands that statement on its head and asserts that the way to end monopoly capitalism, imperialism and other social injustice, is to end war. This is what A.J. Muste meant with his statement, "There is no way to peace; peace is the way."

It would be possible to end racial and sexual discrimination, poverty and other systems of injustice while still continuing the war system. The rival systems of armies, navies, air forces and weapons programs destroy the international fabric of peace, as well as the internal fabric of justice. As Robert McIver indicated, "in the last resort, the cause of institutionalized behaviour is the institution that sanctions it." A UN [United Nations] study revealed that at least fifty million people are directly or indirectly engaged institutionally in military activities worldwide, including about 500,000 scientists and engineers engaged in military research and development, as well as millions in government bureaucracies and weapons production.

National Security Through Armaments

A third weakness is the acceptance by anti-war groups of arms control instead of disarmament as the context for their action. Arms control accepts the idea of national security through armaments. Arms control is the theory that within the armaments system it is possible to reduce the costs of the arms race by international agreements and by certain unilateral actions. Disarmament, on the other hand, is the reduction and/or abolition of arms and military institutions.

The nuclear-freeze movement, which captured the energy of many pacifists, worked to maintain rather than expand the already existing high level of nuclear weapons. Its focus on arms control did not include abolition of the CIA's [Central Intelligence Agency] paramilitary activity—its more than fifty CIA operations around the world through which the US

John M. Swomley, "Where the Disarmament Movement Is Today," *Fellowship*, January/February 1990. Reprinted with permission.

was actually waging imperialist war. The chief contribution of the Freeze movement to disarmament was its advocacy of a nuclear test ban. If you can't test weapons, you can't produce and deploy them with any certainty. Yet neither the Freeze nor any other peace organization focused strongly on unilateral ending of US testing when the Soviets took that unilateral action for eighteen months.

"The world can either continue to pursue the arms race or move consciously to disarm and build an economically and ecologically secure world."

If disarmament is to become a politically significant issue in Washington and in the United States, it must become a central focus of the peace movement. That would require not only a coalition of peace organizations, but building a coalition of other national organizations, whose first task would be changing the mindset of the American people. A militarized mind accepts the proposals of the warmakers, whether that is high military budgets, secret appropriations for the CIA, thousands of overseas military bases, ROTC [Reserve Officers Training Corps] in high schools and colleges, national service or other military demands. A demilitarized or disarmed mind and spirit is a prerequisite for any disarmament program.

A second focus of those concerned with disarmament is to analyze and answer the propaganda and ideology of the war system. Deterrence, for example, is a great hoax or deception. It assumes that an adversary will not attack if the US has sufficient weaponry to inflict unacceptable damage in return. It offers no security, because it is impossible to discern the difference between weapons that are for deterrence and those to be used for a preemptive strike. The damage to the attacking nation would also be unacceptable and counterproductive to the US. Deterrence is a doctrine used to support imperialism and dominance by focusing attention on the USSR and diverting it from Central America and the Middle East.

Securing Verification

Similarly, verification was an obstacle to disarmament, not a means of securing it, since those who refused to negotiate disarmament because it could not be verified spent billions of dollars to develop weapons that could not be verified. They seemed able, however, to specify violations of agreements by their adversary without on-the-spot verification, because apparently they did have adequate information. Verification was an excuse.

Third, new as well as old arguments must be developed for total disarmament. For example, the idea that widespread knowledge of how to develop nuclear weapons makes it impossible to eliminate them also makes it essential to eliminate the arms system and an economy based on it. Even crude weapons cannot be easily produced outside of established production "factories" and cannot be delivered without complex delivery systems such as missiles and bombers.

We must also stress the fact that the world can either continue to pursue the arms race or move consciously to disarm and build an economically and ecologically secure world; it cannot do both. Non-military challenges to national and world security are at least as grave a threat as military dangers. By the same token, we must not acquiesce in the attempt to provide military answers to non-military challenges, as Congress and the Administration have done to thwart production of drugs in Latin America, and their importation to the US.

Economic conversion of the arms industry must also be explored and advocated if we are to provide a decent existence for human life, as well as survival of all life on the planet.

Unilateral Action

Finally, we must recognize the value of unilateral action. It has been clearly demonstrated on the international scene by the Soviet Union—in reduction of military budgets, withdrawal of troops and weapons from other countries, reduction of military personnel and the ending of control over allied and client states. A Soviet official, in a letter to *The New York Times*, referred to their 1987 unilateral action with these dramatic words: ". . . we have a 'secret weapon' that will work almost regardless of the American response—we would deprive America of The Enemy. And how would you justify without it the military expenditures that bleed the American economy white, a policy that draws America into dangerous adventures overseas and drives wedges between the United States and its allies, not to mention the loss of American influence on neutral countries?"

The US can and must unilaterally cut military budgets, reduce the size of its armed forces, stop efforts to control other nations and end all paramilitary and low-intensity conflict in other countries. An essential place for us to focus is the CIA, with its more than $25 billion annual budget, its covert operations, and the entire secrecy surrounding its budget, its personnel, its operations. It has destroyed democracy and justice in government because its personnel cannot be prosecuted for serious crimes. Neither the press nor members of Congress may reveal classified information, and the President can use the CIA to engage in secret wars around the world as if it were his personal private army.

Advocacy of unilateral action can take place in

many ways. It can begin in nonviolent demonstrations, boycott of colleges that permit CIA recruiting, election campaigns against Congressional supporters of subversive CIA activities, legislative action, and education about the value and effectiveness of other unilateral actions which the US could take.

The words "unilateral disarmament," which were formerly smear words designed to demonstrate that its advocates were unrealistic utopians, are now proven examples of realism. It is unilateral diplomacy, unilateral reduction of arms and other similar steps that have ended the cold war, Soviet imperialism, and the threat of invasion of western Europe.

American Imperialism

However, the world is not safe from American imperialism, or from thousands of existing nuclear weapons in the arsenals of the US, Britain, France, China, the Soviet Union, Israel and South Africa. Nor is it safe from chemical and conventional weapons, as numerous wars in Africa, Asia, Latin America and the Middle East in this decade attest. Only a disarmed world can bring security and survival.

John M. Swomley is a former executive secretary of the Fellowship of Reconciliation, a peace organization in Nyack, New York. He is the author of Religious Liberty and the Secular State.

"Properly redefined, the Stealth bomber might become the perfect procurement expense for the winding down of the Cold War."

viewpoint **22**

The Stealth Bomber Will Improve U.S. Defense

Gregg Easterbrook

Defense Secretary Richard B. Cheney, under substantial pressure to cut the defense budget, indicated he would strive to preserve one of the most costly new acquisition programs—the Stealth bomber. Since international tensions are lowering and a huge amount of money—$70 billion or more—is involved in the B-2, Capitol Hill observers had expected the bomber to be an obvious candidate for sacrifice.

Don't look now, but Cheney may be doing the right thing. The B-2, oversold and hard to justify on the Pentagon's preferred terms—as an ultimate wonder weapon for an era of strategic overkill—may become much more attractive with the dawn of *glasnost* and the opening of the Eastern Bloc.

The Stealth bomber has been pilloried by critics as less proficient in doomsday matters than other proposed weapons systems. That's correct—and may make the plane perfect for the 1990s. Whatever their technical specifications, nuclear bombers are simply less threatening than nuclear missiles. We may be bold to hope that, in the 1990s, less proficient doomsday systems will be the kind all major powers switch to. Suddenly, the B-2 may sound like a great idea.

In fact the relaxation of international tensions may do for the B-2 what no Pentagon spokesman has been able to—provide a rationale for its existence. Look at the history of manned bombers throughout the nuclear era:

Because of the tasks assigned modern bombers, discussions about them tend to be gloomy. The bomber pilots' professional paradigm is the execution of a plan for widespread destruction: of supply depots staffed by teen-age conscripts, of factories run by noncombatants or, in the case of nuclear combat, of innocent multitudes. Whatever one may think of the military mind, no God-fearing family man draws satisfaction from this work.

Yet even after the nuclear bomb was invented, strategic bombers seemed like a great idea to military planners. During the decade following the Korean War, when the Pentagon's budget was lower in real terms than today, the Air Force was able to complete five jet bomber construction programs—the B-47, B-52, B-57, B-58 and B-66—turning out several thousand in total. (A maximum of 132 B-2s is planned.) Since B-52 production ended in 1962, the United States has managed to field just 100 more bombers, B-1s, before production ended in 1988.

No Lack of Spending

This has not been for lack of spending—many billions have been invested in failed bomber projects. The problem faced by manned bomber advocates of the Strategic Air Command (SAC) is that the kind of aircraft they long for—winged wonders able to fly halfway around the world, weave through Soviet air defenses protected only by electronic black boxes, then bring their pilots home under Armageddon conditions—don't make sense in a missile era.

What is a nuclear bomber supposed to do that a missile can't do faster, cheaper and without risk to crew? This question, dating back to the Kennedy era, seems more compelling each year as technical advances in electronics and manufacturing make missiles steadily cheaper, while complexities needed to respond to missile improvements make manned aircraft more expensive.

In the swirl of the Washington expert set, debates about whether missiles have rendered the bomber archaic usually turn on convoluted scenarios. There are simpler ways of looking at it. For instance, bombers can be shot down and currently missiles cannot. During World War II, thousands of allied bombers were downed despite the absence of guided antiaircraft weapons and the presence of fighter escorts. Today the Soviet home air defense system has

Gregg Easterbrook, "The B-2 Stealth Bomber: Finding a Purpose Worth a $70-Billion Price Tag," *Los Angeles Times,* December 24, 1989. Reprinted with permission.

approximately 300 SAMs [surface-to-air missiles] per U.S. strategic bomber.

And World War II bombers, though slower than current designs, were bristling with guns for self-defense. The unlikelihood of destroying an approaching SAM missile with a gun, coupled with the need to trim off every pound of weight to make longer flights possible, caused U.S. designers gradually to remove guns from bombers. The B-52, with one cannon set in its tail, was the last "armed" U.S. bomber. Relative to Soviet fighters, the B-1 and B-2 are as defenseless as Korean Air flight 007.

The Tail Gun Factor

This "tail gun factor" has long given many in the Air Force pause. How is a bomber to defeat interceptors over enemy territory? The most far-out plan involved the B-50, a gigantic bomber of the early nuclear era. At one point this plane was to carry in its weapons bay a miniature one-man jet fighter not much bigger than a Dodge minivan. When enemy interceptors appeared, the mini-jet would be dropped into the air and, after fighting off the attackers, fly back to the B-50 for a mid-air recovery in a sling. Right. About the the only way a Pentagon planner believes U.S. bombers will transit Soviet airspace in one piece is by assuming that on-board electronic countermeasure boxes, designed to jam enemy sensors, will work perfectly. But perfection in electronic counter-measures has rarely occurred. Many of the B-1's operational troubles stem from its primary black box. The Air Force admits it may not be able to mask the B-1's presence. It has, however, been known to jam the B-1's own radar.

The Defense Lobby

If some of today's defense-lobby pleas for the B-2 sound detached from reality, it is important to know that, since the late 1950s, all advanced bombers contemplated by the Air Force have been improbable —because all have been attempts to justify why, at great cost, men should go on what are inherently one-way missions. A quick review:

—The initial superbomber was the B-58 Hustler. Operational in 1959, the B-58 was the first heavy supersonic bomber, billed as the wave of the future. Instead B-58s achieved the distinction of being mothballed a few years after leaving the factory. The aircraft burned so much fuel they could barely get to the U.S. border, let alone the Soviet Union's.

—Next came the nuclear-powered bomber. More than $5 billion (1989 dollars) was spent on a conjectural reactor-propelled aircraft supposedly able to remain aloft days at a time. Yet routine calculations showed that even if nuclear engines worked, they would require so many pounds of shielding the plane would not be able to take off. No machine was built and most engineers now consider the whole notion a crackpot idea.

—Concurrently the X-20 Dynasoar, a sort of space bomber, was researched. Resembling a one-man space shuttle, Dynasoar was supposed to be launched into orbit atop a Titan missile. The craft would release a warhead in space, reenter the atmosphere and glide back to a runway under pilot control. Essentially an ICBM [intercontinental ballistic missile] with a man strapped on.

—By the early 1960s, SAC's fondest hopes for sustaining the manned bomber in a missile age centered on the B-70, a proposed superbomber resembling the Concorde, that was to compete with the speed of ICBMs by flying at three times the speed of sound all the way to the Soviet Union. Fuel consumption rates, unfortunately, showed this to be impossible. Ultimately the program met an ignoble end when a B-70 crashed during filming of promotional photographs.

Perennially the Air Force is accused of falling under the spell of hardware: craving top speed and sleekness while neglecting practical considerations of performance at the lesser velocities and altitudes where most military aviation occurs. The $530 million B-2, however, is subsonic. At last the Air Force has found a slow airplane it can love—because it's finally figured out a way to make one incredibly expensive.

"No one doubts the B-2 will be more elusive than the B-1 or the radar-friendly B-52."

The two-decade cessation of bomber production not only indicted the tenuous nature, in the intercontinental ballistic missile age, of arguments supporting manned strategic delivery systems. It signaled a fundamental shift in U.S. military priorities, away from ability to wage conventional war.

Conventional War

Bombing remains applicable for conventional war, but for this purpose aircraft with winged-wonder credentials may backfire, because they cannot be obtained in militarily significant numbers. The U.S. manufactured some 35,000 bombers during the World War II period: far smaller production runs are one reason inflation-adjusted costs per bomber have risen so dramatically. Considering that massive allied bombing failed to halt German war production, or that six times the World War II tonnage—dropped with greater accuracy—failed to overcome North Vietnam, it's hard to imagine how today's relatively tiny contingent of superbombers could do more than incidental damage in a serious conventional war.

In this regard it's important to consider Air Force contentions that the B-2 could be used for conventional war. Stealth works only against electronic detection. If its mission were conventional war in Europe—a much smaller area jam-packed with soldiers who can see a Stealth bomber go by—electronic ghostliness is only a partial help.

No one doubts the B-2 will be more elusive than the B-1 or the radar-friendly B-52. Yet even with the first B-2 now flying, the Pentagon is saying it couldn't dream of shining a radar on the plane till 1991 at the earliest. Congressional suggestions that the Stealth bomber be flown cross-country to determine if it shows up on air traffic control radars have been met with huffy Air Force comments about the primitive technology of civilian tracking systems. One can't help wondering if the generals fear the B-2 would make a cameo appearance on some overworked controller's screen. That would be the end of a $70 billion acquisition right there.

Justifying the Bomber

But given the B-2's breathtaking price tag, at $530 million per aircraft, even radar-evasion properties that work do not alter the case against manned "penetrator" bombers. In technospeak a penetrator flies close to its targets and delivers nuclear bombs via free fall.

One theoretical advantage originally used to justify investment in a "penetrating" Stealth bomber was that such a plane could operate, unobserved, above the Soviet Union at high altitude, away from the pilot stress and physical dangers of treetop flight, beyond the range of ground guns and small SAM missiles. At high altitude, the airplane's flying-wing shape, a handicap during low flight and maneuvering, becomes ideal for fuel efficiency, theoretically giving a Stealth bomber the potential to inspect a large area of the Soviet Union before deciding what to destroy. But in 1988, Air Force Chief of Staff Gen. Larry D. Welch announced that one reason for B-2 cost increases is redesign of the bomber for treetop altitude penetration. This is hardly reassuring: An aircraft on which no expense was spared to reduce detectability may have to hug the ground to escape detection.

Another possible advantage of a Stealth bomber is the ability to go after movable targets that ICBMs, whose destinations are fixed at launch, have missed. Welch has described this as a principal justification for the B-2.

"Another possible advantage of a Stealth bomber is the ability to go after movable targets that ICBMs . . . have missed."

Assuming B-2s were used as Welch suggests, about 10 hours after the initial nuclear exchanges they would scour the Soviet Union for anything still moving, and wipe it out. A gruesome new level of unthinkability? Not necessarily, at least under the old set of assumptions about the evil Soviet Union. Soviet leaders underground in protective bunkers might tell themselves they could survive a missile attack. If they knew a special weapon would be coming later to get them personally, they might be less prone to act rashly.

Chasing relocatable targets is probably the best pure military argument for the B-2. Yet the same task might be executed by cruise missiles, which could be made capable of receiving target updates in flight, using spy satellite information as their guides.

What Future Threat Exists?

While interviewing U.S. fighter pilots recently, I asked what future aerial threat they most feared. Several mentioned the possibility that the Soviets would build comparatively cheap bombers designed to launch Stealth cruise missiles from off our shores: once the little missiles were away, the pilots said, there would be no finding them. Trying to sound like a B-2 salesman I countered, "Don't you fear that the Soviets will copy us, concentrating their resources on a few superbombers designed to fly deep into the United States undetected?" Every fighter pilot rolled his eyes as if someone had let Howdy Doody into the room.

Then are there any reasonable arguments in favor of continuing U.S. possession of manned nuclear aircraft?

Yes, principally that bombers are inherently less threatening than ICBMs. They travel to the target much more slowly than missiles—and can be shot down using current technology. And as SAC officers never tire of pointing out, manned aircraft can be recalled once launched. For these reasons both sides in the Strategic Arms Reduction Talks have concentrated their energies on cutting back missiles, not bombers.

The Strategic Triad

Neither valid reason for manned bombers, however, requires "penetrator" systems as expensive as the B-2. An attractive alternative would be a fleet of simpler bombers intended principally to ferry the advanced cruise missile to a standoff launch point.

Increased reliance on standoff bombers carrying cruise missiles would have nothing to do with eliminating the airborne "leg" of the strategic triad—pilots would still fly the cruise missile carriers near the enemy borders. Such a fleet would preserve the deterrence and stabilizing advantages of bombers while eliminating the huge expense and suicide logic of insisting that a man come within view of the target and toggle a bomb release.

Today Air Force leaders who privately concede that a lower-cost standoff bomber might have been preferable to a B-2 maintain that the point is moot because the B-2 exists and its development costs—about one-third of the price—are sunk and cannot be recovered.

"The B-2 might become a good buy if the Pentagon is willing to cancel some other new strategic systems in return."

But if the Pentagon changes the reasoning for building the B-2, elected officials should not hesitate to change the mission. The B-2 was conceived in a period of rising nuclear tensions. Today there is cause to hope that the Soviet Union will become much less a threat to its neighbors, perhaps even a tenuous U.S. ally.

On the other hand, because the B-2 is inherently less threatening than other strategic delivery systems, its acquisition for use as a cruise-missile carrier might be easier to justify today than before the Berlin Wall fell. If the B-2 were reconfigured as a cruise-missile carrier it could be slightly deglorified technically, lowering unit costs.

A Good Buy

The catch: The B-2 might become a good buy if the Pentagon is willing to cancel some other new strategic systems in return. Ideal candidates are the rail-mobile MX, the truck-mobile Midgetman and the Trident D5 submarine-launched ballistic missile, currently in testing. All are expensive and far more threatening to the world than manned bombers.

It may seem contradictory to argue that a weapon like the B-2 ought to be funded because it is less proficient than available alternatives. But if there is indeed a new golden era of lowered doomsday horizons, less proficient weapons will be the kind in society's best interest. Properly redefined, the Stealth bomber might become the perfect procurement expense for the winding down of the Cold War.

Gregg Easterbrook is a contributing editor for Newsweek, *a weekly newsmagazine.*

"[The B-2] has no military justification; its specific missions are either redundant or actually harmful to US security."

The Stealth Bomber Will Harm U.S. Defense

Art Hobson and Jeffrey Record

Editor's note: The following viewpoint is in two parts. Part I is by Art Hobson. Part II is by Jeffrey Record.

I

The B-2 "Stealth" bomber program, recently given the green light by the Bush administration, is a textbook example of United States defense mistakes. It has no military justification; its specific missions are either redundant or actually harmful to US security. Instead of sound strategic thinking, the B-2 represents a triumph of Air Force and weapons-contract or politics.

The analogy with the MX missile is irresistible. From the standpoint of deterrence, the MX makes no sense because its vulnerability tempts the Soviets to strike first, at the same time that its high "lethality" against Soviet silos pushes them to use their missiles before they are destroyed. But the Air Force wanted more bang for its bucks, so they and their industrial contractors eventually pestered Congress into the first 50 of the 10-warhead MXs, and they may yet get another 50.

Political Punch

The B-2 has yet more political punch. The Air Force likes bombers even more than missiles. Contractors stand to reap even more profits from this $70 billion program—a cool half-billion per airplane—than from the $20 billion MX program. And contractors have taken a lead from the politically successful B-1 bomber program, letting B-2 contracts in all but three states. One company official boasts he has a map of the US in his office, bristling with pins in virtually every congressional district, to show visiting lawmakers.

The B-2's missions will be to penetrate Soviet airspace to deliver nuclear weapons against traditional targets such as military bases and industrial facilities, and secondly to seek out and destroy mobile Soviet missiles. The first mission, penetration, can be performed as well and more cheaply by cruise missiles launched from outside Soviet airspace by our B-52 and B-1 bombers.

The second mission, to destroy mobile missiles, is destabilizing. It is one more example of the fact that, although the Air Force talks a lot about "deterrence," they do not really seem to believe in it. "Deterrence" refers to preventing nuclear war through the threat of retaliation, by either side, for an attack by the other side. In order for this to work, both sides must have forces capable of retaliating. So it makes no sense to build a bomber whose mission is to destroy the very missiles, namely mobile ones, that the Soviets built in order to preserve deterrence!

It was the MX (and the coming Trident II missile) that forced the Soviets into mobile deployments in the first place, because the MX makes fixed Soviet silos vulnerable. The Soviet decision to go to less vulnerable mobiles was admirable. It reinforces deterrence, and it supports the notion that they are not interested in first-strike threats against us, since silo-based missiles are much better for a first strike than are the less accurate and more expensive mobile missiles.

More Harm than Good

And now the Air Force, in its wisdom, wants to target Soviet mobiles, too. It is as though we were bound and determined to force the Soviets into a first-strike posture. Fortunately, the B-2 is not terribly threatening to Soviet mobiles; it will not be effective at destroying them, and the Soviets can choose to launch their remaining mobiles if the B-2 have some effect. But, to the extent that it is successful, the B-2 does more harm than good.

It is a further example of US strategic irrationality that we face a military problem, yet the B-2 makes

Art Hobson, "The U.S. Doesn't Need the Stealth Bomber," *Christian Science Monitor,* June 30, 1989. Reprinted with permission. Jeffrey Record, "Our Disappearing Bomber Force," *Los Angeles Times,* June 8, 1989. Reprinted with permission.

this problem worse while absorbing the funds that could be used to solve the real problem. The real problem is the vulnerability of US MX and Minuteman missiles.

There are solutions: The mobile "Midgetman" missile (half as expensive as the B-2), the shell-game "multiple-silos" basing scheme, "superhard" silos, and getting rid of US land-based missiles altogether as part of an arms control deal with the Soviets. Any of these approaches makes eminent sense.

But what is the US doing? Going ahead with MXs and B-2s that fly in the face of everything we know and profess about the deterrence of nuclear war.

II

In the late 1950s, the U.S. long-range bomber force consisted of 1,854 aircraft, none of them more than seven years old. Today, the force counts only 359 planes, 262 of them B-52s, the last of which rolled off the assembly line 27 years ago. At some point around the year 2000, as the last B-52s are retired, the force could shrink to fewer than 100 aircraft.

Alternatives to Manned Bombers

A decline in the size of the bomber force over the last 30 years was inevitable. The emergence of the intercontinental ballistic missile in the late 1950s and early '60s offered a relatively cheap and, in many respects, more efficient alternative to the manned bomber as a means of delivering strategic nuclear strikes. The almost exponential growth in the cost of building bombers capable of penetrating ever more sophisticated air defenses also made the planes unaffordable in large numbers. Compare, for example, the $800,000 price tag of the B-29, by far the most expensive and technologically advanced bomber of World War II, to the estimated $572-million cost of the Air Force's new B-2 stealth bomber.

Cost growth has reinforced the oft-heard, albeit utterly mistaken, view that the long-range manned bomber has become as obsolete as horse cavalry.

Long-range bombers remain an indispensable component of U.S. security. They add redundancy to the strategic deterrent, and the very existence of a U.S. long-range bomber force vastly complicates Soviet strategic planning. Bombers also contribute significantly to strategic stability, since their relatively slow speed precludes their employment in a first-strike role. Bombers, unlike ICBMs, are also recallable and reusable, and they can deliver large amounts of all kinds of munitions, conventional as well as nuclear.

The importance of this last attribute has all too often been ignored. Since Nagasaki, U.S. long-range bombers have been employed in combat only in non-nuclear conflicts in the Third World (the Korean and Vietnam wars). And conventional strategic bombardment requirements are likely to grow in the future. On the one hand are mounting threats to U.S.

security interests in the Third World. On the other is the continuing contraction in the U.S. overseas basing network, which places a premium on intercontinental range over tactical air power. (The Air Force played no role in U.S. military operations in the Persian Gulf in 1987-1988 because the political precondition for its participation—access to bases ashore in the region—was absent.)

A Bomber Force

Concern over the decline in the size of the U.S. long-range bomber force is therefore legitimate. The United States needs a modern bomber force not only larger than the one now planned (100 B-1Bs and 132 B-2s), but also flexible enough to handle both strategic nuclear as well as conventional contingencies. Unfortunately, it has—and will have—neither, unless present plans are changed.

The principal obstacle to the kind of bomber force that is needed is, ironically, the B-2 itself. Even in the absence of an unprecedented defense budgetary crisis, the B-2 program's cost alone ($75 billion and climbing) would severely limit the size of the future force. Indeed, the program's cost overruns to date have convinced many that it will fall significantly short of its declared goal of 132 planes.

Already twice as expensive as the B-1B, the B-2 is so specialized for its declared mission—namely, to evade Soviet radar detection and seek out and destroy mobile ICBMs and other moving targets deep inside Soviet territory after an initial U.S.-Soviet strategic nuclear-missile exchange—as to make it a very poor candidate for non-nuclear contingencies not involving the Soviet Union. The B-2 is simply too expensive and inflexible to be risked in other than all-out nuclear war.

"The emergence of the intercontinental ballistic missile in the late 1950s and early '60s offered a relatively cheap . . . alternative to the manned bomber."

Overshadowing issues of affordability and strategic utility is whether the plane will perform to expectations. The technological difficulties that have attended the B-1B program (similar in scope to those that initially plagued the B-29) pale in comparison to the ones that hover ominously over the B-2. Technologically, the B-2 is vastly more exotic than the B-1B, which was launched on a solid foundation of almost 2,000 hours of flight-testing experience afforded by the previously canceled B-1A program. Worse still, the B-2 is being developed, tested and produced concurrently rather than sequentially. This concurrency approach would risk program failure for a weapon system far less complex; it begs for disaster in the case of the B-2, which has already suffered slippages in schedule due to technological difficulties.

The B-2's Cost

The B-2's cost, dubious mission and questionable performance argue strongly for its cancellation. In May 1989, Robert Costello, outgoing undersecretary of defense for acquisition, publicly called for the B-2's termination, citing the program's exorbitant cost and persistent contractor mismanagement. Secretary of Defense Dick Cheney has imposed a one-year delay in the program in order to assess more fully its cost, developmental risks and strategic potential. Influential members of Congress, including two prominent Republicans on the Senate Armed Services Committee, are privately on the verge of calling for its abandonment.

"The B-2's cost, dubious mission and questionable performance argue strongly for its cancellation."

The B-2 has become the Air Force's Great White Whale, threatening to drag down scarce dollars that could be much better spent. Cancellation of the B-2, however, will condemn the United States to a modern bomber force of only 97 planes, assuming no more accidental B-1B losses. So small a force would be pitifully inadequate to carry out its share of the strategic nuclear retaliation mission.

What is needed is a new bomber, technologically less fancy (and therefore cheaper) than the B-2 and capable, like the B-1B and modernized B-52G, of continuing to meet the very kinds of non-nuclear long-range bombardment requirements that the Strategic Air Command has actually been called on to perform since 1945.

Since the beginning of World War II, the United States has relied heavily on bombers for deterrence and defense. Can we now allow a situation to unfold in which the number of our long-range bombers in active service is exceeded by that on display at museums?

Art Hobson is a professor of physics at the University of Arkansas in Fayetteville, and is the co-author of The Future of Land-Based Strategic Missiles, *a study done by the Forum on Physics and Society, a division of the American Physical Society. Jeffrey Record is a senior research fellow at the Hudson Institute, a public policy research center in Indianapolis, Indiana. His book is* Beyond Military Reform, American Defense Dilemmas.

"The U.S. Air Force's revolutionary Stealth bomber has the potential to force the Soviets into a debilitating spending binge to create new defenses."

viewpoint 24

The Stealth Bomber Should Be Funded

William B. Scott

While the U.S. Air Force's revolutionary Stealth bomber has the potential to force the Soviets into a debilitating spending binge to create new defenses, basic questions about the B-2's effectiveness and cost must still be resolved.

The B-2 Stealth bomber is technology's answer to the black arts—survival through technological "tricks" that allow an aircraft to hide in the open sky.

From the first design sketch drawn in the late 1970s to the last shot of paint on the B-2's skin in the 1990s, U.S. aerospace expertise has been aimed at one goal: enabling the latest operational bombers to escape the fury of enemy fighters and surface-to-air missiles (SAMs).

Ironically, only by increasing the odds that this sinister looking gray-blue flying wing can reach its target in the Soviet Union and return safely will America guarantee that the aircraft will never have to be used in anger. That's the strange logic of nuclear deterrence.

The Flying Wing

Accomplishing this goal will not be easy. Northrop, designer and builder of the B-2, convinced the Air Force that its 1940s-vintage flying wing design offered the best compromise between small radar cross section and size. Without the large vertical tail and conventional fuselage shape of today's strategic bombers, an "advanced technology bomber" based on a flying wing design reflects less radar energy, yet is big enough to carry 50,000 pounds (22,727 kilograms) of nuclear or conventional weapons. That's less than half the 125,000 pound (56,818-kilogram) payload carried by the more vulnerable B-1B. Judicious contouring of the new bomber's surfaces also minimizes sharp angles that normally reflect radar pulses back to their source.

The Air Force bought Northrop's concept and revived the basic flying wing design. Today's B-2 is almost exactly the same size as its 1940s predecessor, the YB-49, but the similarity ends there. The B-2 stands 17 feet (5.15 meters) high, is 69 feet (20.9 meters) long and has a 172-foot (52.1-meter) wingspan. Although the new bomber looks substantially smaller, its wingspan is close to that of the aging B-52 bomber. The B-2 carries all its weapons internally, and is efficient enough to boast a 6,000-nautical-mile (3,720 kilometers) unrefueled range.

With a single drink from an air refueling tanker, a B-2 can achieve a 10,000-nautical-mile (6,200 kilometers) range, and "can hold virtually every target in the world at risk," according to U.S. Air Force officials. In contrast, the B-1B bomber would require about three times as much tanker support to cover an equal target area.

A key factor affecting the modern flying wing's stealth characteristics—and survivability—has been the evolution of composite materials in recent years. Today, these plastic-like substances are showing up in tennis rackets, cars, modern airliners and next-generation fighters. They are light, strong and terrible radar reflectors, making them ideal candidates for shielding an aircraft from the probing electronic beams of enemy air defenses.

A Functional Aircraft

As a result, the first B-2 built at Northrop's Palmdale, Calif., production plant makes extensive use of radar absorbent composite materials, in combination with complex exterior curves that tend to reflect radar pulses away from their source, rather than back to it. Together, these techniques are expected to make the B-2 virtually invisible to enemy air defense systems.

A head-on look at the bomber reminds one of a cartoon "blob" rising from a pool of gray-blue tar, yet it has a sleek, graceful appearance. The side view is

William B. Scott, "Embattled, 'Invisible' Warrior," *Defense World,* October/November 1989. Reprinted with permission.

another story. The aircraft may be functional, but can hardly be considered attractive—its fuselage is short, and conjures images of a Sea World killer whale.

Featuring wide, wrap around windows that extend almost to the pointed nose, where the left and right wing leading edges intersect, the cockpit area is faired smoothly into the surrounding swept wing surface, eliminating any sharp breaks between wing and fuselage. Low-profile engine inlets that feed air to the two pairs of engines buried in each of the B-2's wings, are placed on both sides of the cockpit, set back several feet from the wing's leading edge. The inlets' outer surfaces also blend into the wing, yet the upper and lower inlet lips are a jagged triangular shape with the points aimed forward. This assures that there are very few surfaces that could reflect an incoming, nose-on radar pulse; only the tiny points of triangles are exposed.

Two doors on top of each inlet provide extra air to the two pairs of engines during high power operation, such as on takeoff and climbout. These doors automatically close in flight, when the aircraft reaches higher speeds.

"The bomber is quite responsive and stable."

The inlets' internal surfaces direct incoming air downward to the four 19,000-pound-thrust-class (8,636 kilogram) General Electric F118-GE100 engines buried in the bomber's thick wing. This location effectively shields the highly reflective compressor face from incoming radar signals The inlet ducts' interiors are treated with radar-absorbent materials as well. These inner surfaces may be fairly delicate and subject to damage during maintenance.

Technologically Complex

A senior Air Force official has acknowledged that design and fabrication of the inlets were a "very technically complex task . . . that took considerably longer than we expected." Some experts question whether this complexity could translate to subsequent headaches for maintenance personnel.

The engines exhaust high velocity hot air over black, heat-resistant composite sections on top of the wing's trailing edge, shielded from heat-sensitive detectors on the ground. Portions of the exhaust area are covered with heat-resistant tiles very similar to those that protect the Space Shuttle during its fiery reentry from earth orbit.

One of the most intriguing aspects of the new bomber is how it is controlled in flight. Northrop and the Air Force are closely guarding details of the B-2's flight control system design, but the first flight provided some insight into its operation. The pilot controls the aircraft by deflecting flaplike surfaces

attached to the sawtooth-shaped trailing edge of the wing. These hinged surfaces move in a complex differential pattern, which causes the aircraft to pitch up and down or roll from side-to-side at the pilot's command.

Yawing the nose to the left or right, without starting an undesired rolling motion, requires separating a split control surface—called a drag rudder—on the outboard ends of the wing trailing edges. Flat surfaces extending at an angle both above and below the left wing, for example, will create enough drag on that side to swing the nose to the left.

Computer Augmentation

A "quad-redundant fly-by-wire" flight control system analogous to that used in the F-16 Fighting Falcon provides computer augmentation for good aircraft stability and responsive handling. Electronic sensors scattered throughout the airframe send a continuous stream of signals to the central computer, which then determines how much to automatically deflect the aerodynamic physical control surfaces, keeping the bomber's nose pointed in the right direction. This technique artificially improves the stability of an aircraft and gives it satisfactory flying qualities.

Test pilots Bruce Hinds of Northrop and Air Force Colonel Richard Couch, who flew the B-2 on its first flight, said the bomber is quite responsive and stable. Hinds said the no-tail flying wing "flies like a real airplane" and is "very nimble."

However, James Kelly, a stability and control engineering consultant in the Los Angeles area, has serious doubts about the B-2's stability. He has questioned the use of so-called "modern control theory" that government and aerospace industry experts have adopted to ensure that the flying wing is stable. Kelly maintains that "classical control theory," which has been the foundation of aircraft designs for years, shows the B-2 is stable in only certain configurations and flight conditions, and could be susceptible to an uncontrollable spin if upset by turbulence or rapid maneuvers.

Northrop and the Air Force claim these stability concerns have been resolved, but the program's classification prevents their explaining just how that is accomplished. Hinds mentioned after the first flight that, although the aircraft handled quite well, minor software changes to the flight control system would be made.

Operational Aircraft

Operational aircraft currently are slated to be flown by a two-man crew, although space has been allocated in the B-2 cockpit for a third member. When the B-2 first flew in July 1989, the Air Force apparently was still undecided whether the crew would consist of two pilots or a pilot and a bombardier/navigator. The 31st Test and Evaluation Squadron at Edwards Air Force

Base, Calif. (where the B-2 will be flight tested), has nonpilot "bomb/navs" assigned to it, but there is no assurance they will be part of the final bomber crew complement. Testing will determine whether two people in the cockpit can handle all the necessary tasks. If not, the third crew position could be activated.

The B-2 normally will be flown by the left-seat pilot, who flies the aircraft with a control stick, similar to that used in fighters, in his right hand, and a set of throttles in his left. The F-111 also uses this control configuration. On the instrument panel, TV-like cathode ray tubes dominate the primary real estate in front of both pilots. Round dials are probably limited to emergency, standby gauges. Airspeed, altitude, heading, attitude and even engine parameters are presented on the TV color displays.

A heavy reliance on built-in computers enables a two-man crew to monitor and control all aircraft systems while still concentrating on its primary mission—directing the aircraft to a target and returning safely.

For low-level penetration missions at 100-to-200-foot (30-60 meters) altitudes, a "covert" terrain following radar will guide the aircraft over and around mountains, buildings, trees and other obstacles. This will allow the bomber to hug the ground, even at night and in bad weather. Other design features will keep infrared and electronic "signatures" to a minimum. Unlike the B-1B, the B-2 will not rely heavily on sophisticated electronic countermeasures to jam enemy defenses; the new bomber is designed to remain virtually invisible to them until it gets within a few miles of a search radar.

Stealth-Related Features

All of the B-2's stealth-related features, though, ultimately come down to how well the aircraft is designed and built. If Northrop and its team of subcontractors do their job right, operational bombers will not show up on radar or infrared detection systems until they are too close for the enemy to do anything about them.

That means taking great pains at every turn of the development and fabrication process. For example, parts have to fit together tightly to make sure that the external skin is absolutely free of gaps and tiny discontinuities that could reflect a radar pulse. To accomplish this, Northrop installed a three-dimensional computer graphics system that links the design, fabrication, assembly, inspection and testing processes for better quality control at less cost than is feasible using conventional paper-based techniques.

This system made possible very difficult tasks, such as fabricating large composite and metal sheets within extremely tight tolerances. Northrop production executives have said much of what their people are doing routinely with the 3-D graphics system could not have been done with conventional techniques.

Entirely new tools and processes had to be developed for the aircraft's construction as well. For instance, Northrop asked Rockwell subsidiary Allen-Bradley to develop a computer-controlled drill that would automatically sense drilling forces and adjust cutting speeds when going from one material into another.

The Bomber's Structure

Because the bomber's structure—in some places—is a sandwich of graphite composite material, titanium and aluminum that varies in thickness and curvature, a high-tech "adaptive drill" was needed for precise holes that could be drilled at economical speeds. The new tool is considered three times more productive than a standard drill and is less likely to cut distorted holes. In some cases, a single badly distorted hole could force a $20,000 sheet of material to be discarded, making the new drill a virtual necessity for economical large-scale production.

"The new bomber is designed to remain virtually invisible . . . until it gets within a few miles of a search radar."

Even riveting or fastening aircraft surfaces together is a carefully controlled, labor-intensive process. First, the hole must be drilled at a precise angle, with the correct shape and dimensions, then be coated with an adhesive and corrosion-control substance. The fastener must be installed according to specification, and its head covered with another coating to create a smooth, nonreflective surface. When finished, the painted skin surface normally will show no discontinuities to identify the fastener's location.

Although Northrop apparently is succeeding at introducing high-precision, computer-based design and manufacturing techniques into the B-2 program, it has not been without a few problems.

The electrical system and wing leading edges have been particularly difficult technical areas. Electrical "harnesses"—bundles of wires bound together into thick cables—are designed, refined and fabricated with computer techniques. But they have often turned out to be too short when they were installed in the aircraft. Other harnesses simply failed to convey power to the right boxes and switches and had to be ripped out and completely reconstructed.

Classified Designs

The wing leading edge designs are classified but are believed to be a complex labyrinth of chambers behind a nonmetallic skin. Radar pulses penetrating the skin could be bounced around until their energy is largely dissipated, ensuring that any reflected signals would be very weak. These, too, have caused engineers and production workers substantial grief.

Dozens of other new techniques and processes have gone into the B-2, all in the name of minimizing the bomber's detectability. Some are sophisticated, others are more mundane, but the Air Force, Northrop and subcontractors are not talking about them—or anything else.

"Maybe the Darth Vader of bombers never will have to fly that mission in anger."

Security on the program has been tighter than that of any development effort since the Manhattan Project brought the U.S. into the nuclear age in the 1940s.

Until 1988, the Air Force would not even admit that the project even existed. Northrop hired thousands of people to work at its Pico Rivera plant in the Los Angeles area, yet would not publicly admit it was for the bomber program. Telephone calls to the Pico facility were answered with a simple "Hello." Unless the caller could give a desired extension, that was the end of the conversation.

Site 4

A huge multimillion-dollar hangar complex Northrop built at Palmdale's airport (on Air Force Plant 42 property) was referred to only as "Site 4." A similar complex built south of the Edwards AFB [air force base] main runway for B-2 testing officially did not exist. When asked what the new buildings were for, Air Force officers would deadpan, "What buildings?"

American taxpayers and members of Congress, although accustomed to weapons systems costing billions of dollars, are suffering from "sticker shock" on this one, and are watching its development closely. While grudgingly agreeing to support the program for at least another year, lawmakers will have a tough time fitting the B-2 massive costs into tight federal budgets over the next few years.

But, if in the late 1990s the B-2 is ever called on to fly "in anger," a two-man crew all alone in the cockpit of the new bomber, hugging the ground at 500 knots over a darkened, unfamiliar countryside, will consider every dime well spent if the B-2 can slip in undetected, hit its target, then get home safely. If it can demonstrate in peacetime that it can meet that goal with absolute certainty, maybe the Darth Vader of bombers never will have to fly that mission in anger.

William B. Scott is the senior engineering editor for Aviation Week & Space Technology. *He is also a flight test engineer in California.*

The Stealth Bomber Should Not Be Funded

Michael Brower

The stealth bomber—whose revolutionary design is supposed to enable it to maneuver undetected inside the Soviet Union—was one of the least controversial of the Pentagon's big-budget programs. While Reagan administration officials and members of Congress butted heads over the MX and Midgetman missiles, the B-2 project breezed through hearings virtually unopposed. But today full-scale production of the bomber appears certain to be delayed, and some influential members of Congress, including Senate Armed Services Committee chair Sam Nunn (D-Ga.), have suggested that the program may have to be canceled.

This turnaround should come as no surprise, even though the bomber has long enjoyed support from Democrats and Republicans alike. The aircraft's enormous cost was bound to raise problems as Congress and the Pentagon struggled to set priorities in a deficit-constrained military budget. The Defense Department now estimates that 132 stealth bombers will cost $68.1 billion. At about $516 million apiece, that is almost twice the price of the B-1B bomber, which was deployed in 1986. Well-publicized difficulties with the B-1B have only highlighted questions about the Air Force's ability to manage such expensive and complex programs.

Underlying the issue of cost is the lack of a compelling need for the stealth bomber. Air-launched cruise missiles—small, low-flying drones that can be launched from aircraft outside Soviet territory—are just as powerful a nuclear deterrent, and they cost far less.

Yet the program will not be easy to kill. The B-2 supporters will be sure to point out that canceling the project could put the prime contractor, Northrop, out of business, along with some 30,000 workers who have already been hired to begin production.

Michael Brower, "In Search of the Elusive Stealth Bomber," *Technology Review*, May/June 1989. Reprinted with permission from *Technology Review*, copyright © 1989.

Moreover, the fact that the program has long been classified leaves little room for debate over its strategic merits—points on which Congress is loathe to challenge the Pentagon anyway. The result could be a compromise all too common with marginal defense projects: a decision to build a smaller number of aircraft over a longer period at an even higher cost.

An Obsolete Role

Despite its extraordinary design, the B-2 bomber has a most traditional strategic role—and one that is all but obsolete. Its mission is to penetrate Soviet defenses and destroy targets at close range, either by dropping free-fall nuclear bombs or by launching nuclear-armed short-range attack missiles (SRAMs). During the 1940s and the 1950s, the penetrating bomber was the only means of delivering nuclear weapons to Soviet soil. Today accurate and reliable long-range missiles have eliminated the need for pilots to fly along with the bombs.

The first challenge to the penetrating bomber's role came in the late 1950s, when President Eisenhower canceled production of a costly new bomber, the B-70, because of the imminent deployment of cheaper and faster intercontinental ballistic missiles (ICBMs). This forced the Air Force to continue using B-52s, which eventually shared equal responsibility with ICBMs and submarine-launched ballistic missiles (SLBMs) in the "strategic triad." In 1976, President Carter halted another attempt to modernize the bomber force by canceling the B-1 program (a decision later reversed by the Reagan administration), this time because of an invention made possible by advancing technology —the cruise missile.

The United States now deploys about 1,500 air-launched cruise missiles (ALCMs) on B-52 bombers. (Despite the age of the B-52s, they are expected to remain in service until after the turn of the century.) ALCMs are inherently difficult for defenses to detect and intercept because of their small size and ground-hugging flight path, and their large numbers can inundate defenses. They are also extremely accurate—exploding within 30 meters of a target—and their 200-kiloton warheads could destroy most "hard" Soviet military sites. And, as with penetrating bombers (but not ballistic missiles), aircraft loaded with cruise missiles can take off from airfields on warning of an attack and then be recalled, without committing the United States to war.

An even more accurate and flexible version of the missile known as the advanced cruise, which makes heavy use of stealth technology, is now being developed. Once deployed on B-1B and B-52 bombers in the 1990s, these missiles should be virtually impossible for Soviet defenses to stop.

Using Cruise Missiles

Cruise missiles not only render the B-2 unnecessary but are also much cheaper. The ALCMs now in service have cost a total of about $4 billion, and the new advanced cruise missiles are expected to cost about $7 billion—one-tenth the price of the B-2 program. Since the missiles would be launched from outside the reach of Soviet defenses there is no need for extremely sophisticated, expensive aircraft to carry them.

Despite the advantages of the cruise missile, the Air Force—long identified with the penetrating bomber—has resisted efforts to fully incorporate it into the arsenal. In the early 1970s, the Air Force insisted on limiting the range of the weapon so that it could not reach all parts of the Soviet Union from outside its borders. The program finally had to be placed under Navy control to ensure that it would be completed. Even now, ALCMs are mostly relegated to the secondary—and wasteful—role of destroying Soviet defenses to clear a path for penetrating bombers.

Planning to Fight a Nuclear War

The Air Force, of course, has outlined specific missions for the B-2 that cannot be handled by cruise missiles, but they are dubious. The most important is to hunt down Soviet mobile ICBMs and command centers (known as "strategic relocatable targets") that might survive an initial nuclear exchange. The Soviet Union began deploying the truck-mounted SS-25 ICBM in 1986, and the rail-borne SS-24 ICBM last year. The Pentagon expects the Soviets to mount up to half their ICBM warheads on mobile launchers by the mid-1990s.

However, it is doubtful that the B-2 would be effective against mobile targets. Trees and rough terrain, simple camouflage, decoys, and the targets' own stealth technology could foil efforts to detect them. Clouds and bad weather would block the B-2's infrared and optical sensors unless the aircraft flew at a very low altitude, where the sensors' range would be limited. And if the bomber used radar to search in any weather, Soviet receivers known as "emitter-locators" could quickly locate the radar's emissions, thus nullifying much of the aircraft's stealthiness.

Just as important, the subsonic B-2 would have no more than a few hours to find its targets, and could not return for more missions, since airfields in the United States, Europe, and Asia would probably not survive a nuclear exchange. The bomber would almost certainly have to rely on information from reconnaissance satellites to narrow its search area quickly. But Soviet antisatellite weapons could destroy U.S. satellites in a conflict, and Soviet transmitters could jam space-based radars with noise.

The Air Force itself all but confirmed this view in November 1988, when chief of staff Gen. Larry D. Welch admitted that mobile targets would be but a small part of the B-2's mission. He added that "the whole business of locating mobile missiles . . . is a very complex task, and we're a long way from having decided that we know how to handle that task." The joint chiefs also retreated in their most recent posture statement, saying that the B-2 would present only "an increased threat to some relocatable targets." Even the Reagan administration's last budget request cut two-thirds of the funds for the Air Force's Relocatable Target Capability Program, which is intended to develop sensors to enable the B-2 to locate mobile targets.

A Marginal Advantage

Air Force officials also claim that the stealth bomber could be effective against "super-hard" targets buried deep underground—principally Soviet leadership and command shelters. The B-2 could deliver a larger bomb with somewhat higher accuracy than missiles, and would thus need fewer weapons to destroy the same number of targets. But this argument collapses on economic grounds: the B-2's enormous cost would more than outweigh the marginal numerical advantage.

"Cruise missiles not only render the B-2 unnecessary but are also much cheaper."

By favoring the B-2 for this mission, the Air Force may actually be trying to take advantage of an arms-control loophole. Conferees at the strategic arms reduction talks (START) in Geneva have agreed to count bombers carrying gravity bombs or short-range missiles as one weapon, but to count bombers carrying cruise missiles as several weapons (probably 10). These artificial rules—a Soviet concession to the United States, which favors bombers—let the Air Force load penetrating bombers with more weapons than it otherwise could. Such an advantage would be especially important under a START treaty that cut the Soviet and U.S. arsenals by 50 percent.

The most disturbing aspect of the B-2's missions is that they have little to do with preventing a nuclear war and much to do with fighting one. Pentagon officials often state that penetrating bombers offer greater flexibility than missiles, allowing the Air Force to respond to changing circumstances. For example, the bombers could ignore targets that had already been destroyed or strike targets missed during an initial nuclear exchange. But these tasks play a role in scenarios of a limited and prolonged war. The proper role of nuclear weapons is to deter war, and for that job ICBMs, SLBMs, and cruise missiles are more than adequate. Indeed, threatening Soviet mobile ICBMs and command shelters with B-2 bombers might only encourage Soviet leaders to fire their missiles early in a conflict to avoid the risk of losing them entirely.

Bankrupting Soviet Defenses

An important goal of B-2 enthusiasts is to make existing Soviet defenses obsolete and force the USSR to invest heavily in new ones, preventing them from building more tanks or missiles. Estimates of the total Soviet investment required to defend against the bomber have ranged as high as $500 billion, though no one has provided evidence to support this figure.

But this reasoning is flawed in several respects. First, U.S. cruise missiles have already rendered existing Soviet defenses largely ineffective—and the advanced cruise missile will make this even more true. Second, the Soviet Union has never tried to make its defenses perfect, and there is no reason to believe that it would bankrupt itself to do so in the future. And third, experience suggests that Soviet spending on defenses is relatively insensitive to changes in the U.S. bomber force. Pentagon figures show that the Soviet budget for operating and modernizing air defenses has remained fairly constant at about $15 billion for 20 years, despite the deployment of cruise missiles and B-1B bombers.

The B-2 might actually prove vulnerable to unconventional defenses that could be cheaper than $500 billion. In fact, the Pentagon is studying several of these as a hedge against future Soviet stealth bombers and cruise missiles. For example, long-wavelength radars, although not very accurate, might work when combined with other radars. (Materials designed to absorb long wavelengths could be too thick and heavy for use in aircraft.) Radars on satellites might also be effective, since no amount of radar-absorbing material could hide the B-2's broad wing when viewed from above. Another possibility is bistatic radar, in which widely scattered ground-based radar transmitters and receivers could detect the aircraft as it flew between them. The Soviet Union might even decide to mount extremely powerful radars and other sensors on dirigibles.

Each of these potential defenses has disadvantages, and the Soviet Union might not want to use them throughout the country. But some might be practical around highly defended sites such as leadership shelters.

If the B-2 program is far from bankrupting the Soviet defense budget, it can be blamed for a significant share of the crisis facing U.S. military spending. The stealth bomber was born in a period of flush budgets and a rapid arms buildup. Today the nation cannot afford to build it without a significant drain on other defense programs. The B-2's 1990 allocation of about $4 billion is about half the requested budget for procuring combat aircraft ($8.4 billion), and one-fourth the requested budget for all aircraft ($18 billion). Strains on Air Force funds have already slowed production of the F-15 fighter aircraft from 48 per year planned three years ago to 36 today, and F-16 production has been cut from 216 to 150 per year.

The price of the B-2 program can be expected to rise in the future: the program has already logged $4 billion in overruns in the R&D [research and development] phase. To make matters worse, the Air Force intends to produce a number of aircraft before testing is fully complete. This policy, called "concurrence," is intended to save money in a low-risk program, but it can be disastrous in a high-risk venture where major problems are likely to crop up. This was the lesson of the B-1B bomber, which began operating before deep flaws were discovered in its electronic defense systems.

"The B-2's enormous cost would more than outweigh the marginal numerical advantage."

The B-2's innovative design, so far unproven, will require extensive testing. For example, the aircraft's unconventional aerodynamic controls (there is no vertical tail or rudder) could make it hard to fly. The "fly-by-wire" computer-control system, needed to compensate for these difficulties, is far more sophisticated than any yet developed. The aircraft is made of composite materials never applied so extensively in such a large aircraft; structural failures could occur. And only full-scale tests will determine whether radar and other sensors can actually detect the bomber.

Northrop's difficulties with the MX guidance system do not inspire confidence that it can manage such complex projects. (About half the deployed MXs have not been working at any given time, and test missiles have had to be cannibalized for parts.) Former employees of Northrop have sued the company for fraud on the B-2, claiming that engineers sat around with nothing to do while the government paid for their time. Robert Costello, undersecretary of defense for acquisition, returned from a plant visit in spring 1988 so disturbed with Northrop management that he proposed canceling the program. The Defense Acquisition Board overruled him, and Pentagon officials now claim that the management issues have been resolved.

Northrop's problems may result from rapid growth. Its annual sales more than tripled between 1980 and 1988, rising from less than $2 billion to about $6 billion. Much of that growth can be attributed to the B-2 program, so canceling or even delaying it would devastate the company. It could also have a severe impact on employment in some communities, since Northrop and its principal subcontractors have already hired over 30,000 people to work on the B-2. This fact is likely to weigh heavily on members of Congress when they vote on the military budget.

Secrecy as Policy

How did the B-2 program reach this stage without significant debate in Congress? The overabundance of military funding during the last decade, and the B-2's identification as a "Democratic" bomber, have contributed, but one of the main reasons is secrecy. Classification has kept information out of the hands of a potentially critical public and limited congressional debate over the bomber.

For much of the 1980s the Pentagon was willing to admit only that the project existed. This secrecy began to lift slightly in 1986, when Secretary of Defense Caspar Weinberger had to give Congress an estimate of the program's total cost—then $36.6 billion. But that figure was in 1981 dollars, and the Pentagon was already aware of substantial cost overruns. More information became available in 1987 and 1988, including an artist's sketch of the bomber and updated (but unofficial) cost estimates, culminating in the unveiling of the bomber in November 1988. But even now such mundane details as the B-2's yearly budget and production schedule remain classified.

"The B-2's innovative design, so far unproven, will require extensive testing."

This pattern is becoming increasingly common for the Air Force. Some 35 percent of its procurement budget is "black," compared with less than 5 percent for the Army and Navy. About 37 percent of the Air Force R&D budget is also classified, compared with 28 percent for the Navy and 17 percent for the Army. Pentagon officials argue that secrecy is needed to protect sensitive technologies, particularly with stealth. (The advanced cruise missile is a black program, for example.) But it is not clear why details such as the programs' budgets should be kept confidential.

A more persuasive explanation is that secrecy is intended to give the B-2 an advantage in budget debates. Until now this strategy has probably been effective, as it has restricted debate to the armed services and appropriations committees, whose members tend to be pro-military and who have little incentive to fight classification (they have clearances). Secrecy has also distracted Pentagon critics, who have focused their attention instead on more visible programs.

Drawbacks to Secrecy

Still, B-2 proponents are finding that secrecy can have drawbacks. The total price of producing the bomber came as a shock to the public even though the actual increase over the 1981 estimate amounted to less than 20 percent after inflation. Secrecy has annoyed the normally pro-military defense press, which has repeatedly called for declassifying the program. And no strong public constituency has formed around the program since the Pentagon has not revealed which districts receive B-2 funding. This is in sharp contrast to the B-1B program, which was famed for having contracts in nearly every state.

Secrecy also carries technical and engineering risks. The B-2 program is heavily compartmentalized, so that one engineering team often may not know what another team is doing, and even senior managers may not fully grasp all aspects of the program. Peer review is limited, and no outside critics keep program managers honest. These drawbacks may have affected the advanced cruise missile, whose serious problems in flight testing have delayed deployment.

The B-2's Fate

The ultimate fate of the B-2 is now anybody's guess. Congress should cancel the program. But the more likely course—one supported by House Armed Services Committee chair Les Aspin (D-Wisc.)—is to stretch out the program and possibly cut the total number of bombers produced. Such a solution could increase the price of each bomber by as much as 50 percent—and add little to our security in the process.

Michael Brower is an arms analyst for the Union of Concerned Scientists, an organization in Boston, Massachusetts, that analyzes the effect of advanced technology on society and advocates peace through arms control.

bibliography

The following bibliography of books, periodicals, and pamphlets is divided into chapter topics for the reader's convenience.

Arms Control

Kenneth L. Adelman — *The Great Universal Embrace.* New York: Simon & Schuster, 1989.

Graham T. Allison and William L. Ury — *Windows of Opportunity.* Cambridge, MA: Ballinger Publishing Co., 1989.

Hugh Beach — "The Case for the Third Zero," *Bulletin of the Atomic Scientists,* December 1989.

James Blackwell — "We Still Need Conventional Arms," *The World & I,* April 1990.

Bulletin of the Atomic Scientists — "Ten Minutes to Midnight," April 1990.

George Bush — "Evolution in Europe," *The New York Times,* May 4, 1990.

George J. Church — "Here We Go, On the Offensive," *Time,* June 12, 1989.

Jonathan Dean — "Can NATO Agree on Arms Control?" *Technology Review,* October 1989.

Jonathan Dean — *Meeting Gorbachev's Challenge: How to Build Down the NATO-Warsaw Pact Confrontation.* New York: St. Martin's Press, 1990.

Damian Durrant and Jacqueline Walsh — "Nuclear Weapons for a Bygone Era," *Bulletin of the Atomic Scientists,* April 1990.

Joshua M. Epstein — *Conventional Force Reductions.* Washington, DC: The Brookings Institution, 1990.

Charles C. Flowerree — "On Tending Arms Control Agreements," *The Washington Quarterly,* Winter 1990. Available from the Center for Strategic and International Studies, 1800 K St. NW, Suite 400, Washington DC 20006.

Michael R. Gordon — "Soviets Rebuffed by Cheney on Plan Curbing Sea Arms," *The New York Times,* April 16, 1990.

Robert Jervis — *The Meaning of the Nuclear Revolution.* Ithaca, NY: Cornell University Press, 1989.

David T. Jones — "How to Negotiate with Gorbachev's Team," *Orbis,* Summer 1989.

Max M. Kampelman — "START: Completing the Task," *The Washington Quarterly,* Summer 1989. Available from the Center for Strategic and International Studies, 1800 K St. NW Suite 400, Washington, DC 20006.

Alexei Kireyev — "Crawling Towards Disarmament," *New Times,* no. 10, March 6-10, 1990.

David Lauter and Norman Kempster — "Bush Asks New A-Arms Cutback," *The New York Times,* May 4, 1990.

Los Angeles Times — "Moscow Has Bad START II Jitters," April 10, 1990.

Morris McCain — *Understanding Arms Control: The Options.* New York: W.W. Norton, 1989.

Robert S. McNamara — *Out of the Cold.* New York: Simon & Schuster, 1989.

Michael Mandelbaum — *The Other Side of the Table: The Soviet Approach to Arms Control.* New York: Council on Foreign Relations, 1990.

John Newhouse — *War and Peace in the Nuclear Age.* New York: Alfred A. Knopf, 1989.

Bernard D. Nossiter — "The Pentagon's Closet Pacifists," *The Progressive,* February 1990.

Ray Perkins, Sidney D. Drell, and Theodore B. Taylor — "Nuclear Abolition: Would Cheaters Count," *Bulletin of the Atomic Scientists,* December 1989.

Edward Rhodes — *Power and MADness.* New York: Columbia University Press, 1989.

C. Paul Robinson and Les Paldy — "Substantial Progress in Nuclear Testing Talks Verifications Protocols Nearing Completion," *NATO Review,* February 1990.

James P. Rubin — "START Finish," *Foreign Policy,* Fall 1989.

Michael Ruhle — "The Illusion of Strategic Stability," *Global Affairs,* Spring 1990.

Amity Shalaes — "Rowny's Warning: Slow Down on START," *The Wall Street Journal,* May 10, 1990.

Randy Shannon — "Time to Renew the Push," *People's Daily World,* October 13, 1989. Available from *People's Daily World,* 239 W. 23rd St., New York, NY 10011.

Baker Spring and Michael Lind — "The State Department and Arms Control," The Heritage Foundation *Backgrounder,* June 1, 1989. Available from The Heritage Foundation, 214 Massachusetts Ave. NE, Washington, DC 20002.

Kenneth W. Thompson — *Richard Garwin on Arms Control.* Lanham, MD: University Press of America and The Miller Center of the University of Virginia, 1989.

Robert C. Toth — "Soviets Likely to Ask Deeper Troop Cuts in Central Europe," *Los Angeles Times,* February 8, 1990.

The Arms Race

Harry Anderson	"Your Move Again, George," *Newsweek,* July 19, 1989.
Joseph Bernardin	"The Changing Nuclear Debate in the 1990s," *Origins,* March 1, 1990.
Bulletin of the Atomic Scientists	"Ten Minutes to Midnight," April 1990.
William F. Buckley	"How to End the Cold War," *National Review,* November 10, 1989.
William F. Buckley	"The Russians Are Still There," *National Review,* December 31, 1989.
Stephen Budiansky	"The Russians Aren't Coming," *U.S. News & World Report,* November 27, 1989.
McGeorge Bundy	"Ending a Common Danger," *The New York Times Magazine,* August 20, 1989.
Arthur C. Clarke	"First Word," *Omni,* October 1989.
Jonathan Dean	*Meeting Gorbachev's Challenge: How to Build Down the NATO-Warsaw Pact Confrontation.* New York: St. Martin's Press, 1990.
William R. Doerner	"A Rush to Sign New Accords," *Time,* February 26, 1990.
Nils Petter Gleditsch and Olav Njolstad	*Arms Races: Technological and Political Dynamics.* Oslo, Norway: International Peace Research Institute, 1990.
Global Affairs	"The Changing U.S.-Soviet Strategic Balance," Spring 1990.
Patrick Glynn	"Nuclear Revisionism," *Commentary,* March 1989.
Michael Howard	"A Farewell to Arms?" *International Affairs,* Summer 1989.
Jonathan Kapstein	"At Ease: The Disarming of Europe," *Business Week,* February 19, 1990.
Michael T. Klare	"Who's Arming Who?" *Technology Review,* May/June 1990.
Jay P. Kosminsky	"America's Stake in the NATO Nuclear Debate," *The Heritage Lectures, No. 219,* July 13, 1989. Available from The Heritage Foundation, 214 Massachusetts Ave. NE, Washington, DC 20002.
Flora Lewis	"Disarm, Don't Disengage," *The New York Times,* September 17, 1989.
John Mueller	*Retreat from Doomsday: The Obsolescence of Major War.* New York: Basic Books, 1989.
John Newhouse	*War and Peace in the Nuclear Age.* New York: Alfred A. Knopf, 1989.
Richard Perle	"The Finer Points of Perestroika," *U.S. News & World Report,* October 2, 1989.
Richard Perle	"If the Warsaw Pact Is Past, Does NATO Have a Future?" *The Heritage Foundation Lectures, No. 236,* December 19, 1989. Available from The Heritage Foundation, 214 Massachusetts Ave. NE, Washington, DC 20002.
William J. Perry	"Defense Investment Strategy," *Foreign Affairs,* Spring 1989.
Dan Quayle	"Strategies for the 1990s," *Global Affairs,* Spring 1990.
Scott D. Sagan	*Moving Targets: Nuclear Strategy and National Security.* Princeton, NJ: Princeton University Press, 1989.
William A. Schwartz and Charles Derber	*The Nuclear Seduction: Why the Arms Race Doesn't Matter—And What Does.* Berkeley: University of California Press, 1990.
John M. Swomley	"U.S. War Momentum Stunned," *The Churchman,* November/December 1989. Available from Promoting Enduring Peace, PO Box 5103, Woodmont, CT 06460.
Bruce Van Voorst	"An Exercise in Trust," *Time,* July 31, 1989.
Bruce Van Voorst	"Reading the Fine Print," *Time,* October 9, 1989.
Sergei Volovyets	"Great Expectations," *Soviet Life,* November 1989.
George Weigel	"The Next Line of Hills: The Challenge of Peace Revisited," *First Things,* April 1990. Available from The Institute on Religion and Public Life, PO Box 3000, Dept. FT, Denville, NJ 07834.
P.A. Woodward	"The 'Game' of Nuclear Strategy: Kavka on Strategic Defense," *Ethics,* April 1989.

Economics of the Arms Race

Bruce B. Auster	"A Healthy Military-Industrial Complex," *U.S. News & World Report,* February 12, 1990.
Bruce B. Auster and Stephen Budiansky	"Why Peace Is So Expensive," *U.S. News & World Report,* March 5, 1990.
John Barry and Rich Thomas	"Getting Ready for Future Wars," *Newsweek,* January 22, 1990.
A. Ernest Fitzgerald	*The Pentagonists: An Insider's View of Waste, Mismanagement, and Fraud in Defense Spending.* Boston: Houghton Mifflin, 1989.
Dave Griffiths	"Cheney Goes to Battle," *Business Week,* February 12, 1990.
Richard Hornik	"The Peace Dividend: Myth and Reality," *Time,* February 12, 1990.
William W. Kaufman	"A Plan to Cut Military Spending in Half," *Bulletin of the Atomic Scientists,* March 1990.
William W. Kaufman	*Glasnost, Perestroika, and U.S. Defense Spending.* Washington, DC: The Brookings Institution, 1990.
Jay P. Kosminsky	"Four Imperatives for Cutting the Defense Budget," The Heritage Foundation *Backgrounder,* March 2, 1990. Available from The Heritage Foundation, 214 Massachusetts Ave. NE, Washington, DC 20002.
Charles Krauthammer	"Don't Cash the Peace Dividend," *Time,* March 25, 1990.
Irving Kristol	"There Is No Military Free Lunch," *The New York Times,* February 2, 1990.
Mark Levinson	"A Peace Dividend for the U.S. Economy?" *Dissent,* Spring 1990.
Thomas L. McNaugher	*New Weapons, Old Politics: America's Military Procurement Muddle.* Washington, DC: The Brookings Institution, 1989.
The Nation	"Time to Pay the Dividend," January 29, 1990.
Sam Nunn	"Cut U.S. Troops Below the 'Floor,'" *Los Angeles Times,* April 20, 1990.
Sam Nunn	"Defense Budget Blanks," *Vital Speeches of the Day,* April 15, 1990.
Philip A. Odeen and Gregory F. Treverton	*Thinking About Defense Spending.* New York: Council on Foreign Relations, 1989.
George Rathjens	"On Cutting the Budget in Half," *Bulletin of the Atomic Scientists,* April 1990.

Robert C. Richardson	"Cutting Defense Budgets Without Guidance," *Conservative Review*, February 1990. Available from the Council for Social and Economic Studies, 1133 13th St. NW, Suite C-2, Washington, DC 20005-4297.
Daniel Seligman	"Pick a Number," *Fortune*, January 1, 1990.
Jed C. Snyder	"Conventional Defense and European Security," *The World & I*, January 1990.
Herbert Stein	"Remembrance of Peace Dividends Past," *The American Enterprise*, March/April 1990. Available from The American Enterprise Institute for Public Policy Research, 1150 17th St. NW, Washington, DC 20036.
Gregory F. Treverton	"The Defense Debate," *Foreign Affairs Special Issue*, 1990.
Bruce Van Voorst	"Sticking to His Guns," *Time*, May 7, 1990.
The Wall Street Journal	"End of the Game," February 14, 1990.
The Wall Street Journal	"A New Strategy?" February 7, 1990.
Murray Weidenbaum	"Defense 'Conversion Plan' Flawed," *Los Angeles Times*, February 11, 1990.
Walter Williams	"Congress Would Spend 'Dividend,'" *Conservative Chronicle*, February 7, 1990. Available from *Conservative Chronicle*, Box 11297, Des Moines, IA 50340-1297.

The New Republic	"Don't B-2 Sure," September 4, 1989.
Bill Sweetman	*Stealth Bomber: Invisible Warplane, Black Budget.* Osceola, WI: Motorbooks International, 1989.
Time	"Bombing Run on Congress," January 8, 1990.
U.S. News & World Report	"Bat Plane Dodges Flak at the Capitol," July 31, 1989.
Bruce Van Voorst	"The Stealth Takes Wing," *Time*, July 31, 1989.
Caspar W. Weinberger	"The B-2—A Low-Cost Insurance Policy," *Forbes*, August 21, 1989.
Elmo R. Zumwalt and Worth H. Bagley	"Technology Is Key to a Strong Defense," *Conservative Chronicle*, December 26, 1989. Available from *Conservative Chronicle*, PO Box 11297, Des Moines, IA 50340-1297.

The Stealth Bomber

D.F. Bond	"USAF Believes Impulse Radar Not Feasible for Detecting B-2," *Aviation Week & Space Technology*, February 26, 1990.
Hodding Carter III	"The Perils of Stealth Is an Old-Time Melodrama," *The Wall Street Journal*, August 3, 1989.
Lee Feinstein	"Heard on the Street," *Common Cause Magazine*, July/August 1989.
Donald E. Fink	"B-2 First Flight," *Aviation Week & Space Technology*, July 24, 1989.
P.A. Gilmartin	"GAO Urges Cuts in B-2 Funding, Production," *Aviation Week & Space Technology*, February 26, 1990.
Alex Gliksman	"Don't Throw the Stealth Out with the Bomber," *The Washington Post National Weekly Edition*, August 14-20, 1989.
Jay H. Goldberg	"The Technology of Stealth," *Technology Review*, May/June 1989.
Melissa Healy	"B-2 Bomber a Visible Target for Budget Ax," *Los Angeles Times*, March 18, 1990.
Melissa Healy	"B-2 Critics Want Cheaper 'B-3' Cruise Missiles," *Los Angeles Times*, April 28, 1990.
Melissa Healy	"Cheney Asks Cuts in B-2, Other Craft to Save $34 Billion," *Los Angeles Times*, April 27, 1990.
Melissa Healy	"Cost of $1.95 Billion for Each B-2 Held Possible," *Los Angeles Times*, April 4, 1990.
David A. Hoekema	"The B-2: Winning Weapon of the Last War," *The Christian Century*, August 30-September 6, 1989.
John Isaacs	"B-2 or Not B-2," *Bulletin of the Atomic Scientists*, September 1989.
Jeffrey Kluger	"Plane Talk," *Discover*, January 1990.

index

Adams, Gordon, 52
Adelman, Kenneth L., 63
Afghanistan
  civil war, 57
  Soviet invasion, 24, 48
Agran, Larry, 34
aircraft carriers, 53, 85, 86
Akhromeyev, Sergei, 78, 84
Altenburg, Wolfgang, 89
Andreotti, Giulio, 12
Angola, 73
anti-ballistic missile systems, 7
  Soviet, 29
  U.S., 3, 49
Anti-Ballistic Missile (ABM) Treaty (1972), 49, 70, 71
  Krasnoyarsk radar violates, 19
  Star Wars violates, 44
    con, 17
arms control, 3, 8, 9, 16-17, 18
  agreements, 91, 94
    can be verified, 19, 72, 89-90
      con, 11-12, 65-66, 73, 86
  conventional force reductions, 22-23, 49, 55, 73-75
    are beneficial, 59-62
      con, 63-68
  naval
    is necessary, 77-81
      con, 83-87
    U.S. attitude toward, 44, 79, 80
  promotes peace, 15, 89-91
    con, 93-95
  Reagan and, 5
  Soviet Union and, 69
  START
    is necessary, 69-72
      con, 73-75
  U.S. and, 7
arms race, 94
  deterrence and, 2-3, 4-5
  is over, 7-9
    con, 11-13
  Soviet Union and, 15
  U.S. perpetuates, 8, 21-26
    con, 15-19
Aschauer, David Alan, 34, 45
Ashoka (emperor of India), 93
Aspin, Les, 52, 53, 54, 80, 112

Baker, James A., III, 15, 73, 74-75, 79
Beall, Donald R., 33
Beatty, Jack, 41
Berlin Wall, 23, 41, 100
Bing, George, 56
Bismarck, Otto von, 62
Blackwell, James, 52
Boeing Company, 33
Bohr, Niels, 4

Bolivia, 44
Bonder, Seth, 54
Boulding, Kenneth, 37
Brezhnev, Leonid I., 24, 51
Brodie, Bernard, 4
Brower, Michael, 109
Brown, Harold, 58, 63-64, 71
Brzezinski, Zbigniew, 39, 72
Bumpers, Dale, 46
Bush, George, 31, 63
  and education spending, 43
  and European troops, 22-23, 24, 51, 75
  and Poland, 43
  and South America, 44
  and Soviet Union, 15, 52, 73
  and Stealth bomber, 101
  arms control policies, 59, 90
    are constructive, 18, 74
      con, 5, 25, 32, 66, 79, 81, 90
    should cut defense spending, 48, 53
      con, 21

Canada
  defense spending, 72
  U.S. and, 68
Canberra conference, 18
Carlucci, Frank, 50
Carnesale, Albert, 60
Carter, Jimmy, 24, 72, 109
  and arms control, 66
Carus, Seth, 57
Carver, Lord, 71
Castro, Fidel, 21
Catherine II (empress of Russia), 83
Catholic bishops, 5, 13, 71
Celeste, Richard F., 34
Center for Defense Information, 27-28
Center for Strategic and International Studies, 33, 56
Chamberlain, Neville, 64
chemical weapons, 16, 18, 57, 61, 64
  in Iran-Iraq War, 13, 48, 86
  prohibition of, 12, 65-66
Cheney, Richard B., 57
  and defense budget, 25, 26, 32, 33, 52-53, 55
  and Stealth bomber, 97, 103
China
  and India, 68
  economy, 38
  nuclear forces, 3
Church, George J., 51
Churchill, Winston S., 2, 5, 19, 65
Clausewitz, Karl von, 2, 15
Clifford, Clark M., 71
Cold War, 28, 31, 41, 46, 62, 67
  and arms race, 7, 8, 54
  and naval weapons, 77, 81
  and Stealth bomber, 100

arms control helped end, 59-62
  con, 63-68
Committee for Economic Development, 43
communism, 4, 93
  in Eastern Europe, 51, 67, 75
  U.S. and, 55-56, 57
Conference on Security and Cooperation in Europe (CSCE), 62
confidence- and security-building measures, 18, 61, 79, 80
conventional forces
  and defense spending, 45, 49, 64
  and nuclear war, 8, 12
  European reduction, 7, 32, 41, 78
    promotes security, 15, 72, 91
      con, 22, 23
  NATO, 3, 45
  Soviet, 48, 59, 61, 67, 90
  U.S., 25, 55
  Warsaw Pact, 45, 89
Conventional Forces in Europe (CFE)
  negotiations, 56, 62, 73
    and NATO, 89, 90, 91
    and Soviet Union, 17, 19, 57, 74-75, 78
Costello, Robert, 103, 111-112
Couch, Richard, 106
Crowe, William J., Jr., 29, 57-58, 65, 85
Cuban missile crisis, 15, 16, 28, 65
Czechoslovakia
  Soviet invasion of, 24
  Soviet troops in, 56, 74

defense contractors
  budget cuts and, 32, 33-34, 54
  Star Wars and, 44
  Stealth bomber and, 101
defense spending, Soviet, 9, 39, 67, 90, 111
defense spending, U.S.
  amounts, 27, 28, 36, 53, 54, 57
  Congress and, 21, 26, 32, 34, 36, 52
  cuts in
    are necessary, 9, 25, 27-28, 72
      con, 29-30
    improved Soviet-American relations justify, 51-54
      con, 55-58
    should fund social programs, 41-46
      con, 47-50
    would improve economy, 8, 31-34, 46
      con, 35-39
  Defense Department and, 30, 31, 37, 52, 53, 54
  on Stealth bomber, 97
democracy
  in Eastern Europe, 24, 25, 43
  in Soviet Union, 16
détente, 1
deterrence, 2-3, 4, 8, 12, 64, 72
  arms control and, 17, 66, 69

defense budgets and, 16
is beneficial, 13, 19
   con, 26, 42, 94
MX missile and, 101
Reagan and, 5
Stealth bomber and, 101-102
U.S. and, 30, 48, 52, 55, 56
Deutch, John, 66
disarmament, 3
Downey, Tom, 42
Drell, Sidney D., 11
DRI/McGraw-Hill, 31, 32, 33
drug war, 44, 55, 87, 94

Easterbrook, Gregg, 97
Eastern Europe, 31, 41, 55
   Cold War and, 62
   democratization, 16, 22, 23-24, 25
   NATO and, 45
   Soviet troops in
      reductions in, 18, 25, 32, 51, 74, 90
      withdrawal of, 22-23, 56, 75
East Germany
   Soviet missiles in, 66
   Soviet troops in, 56, 74
education, 42-43
Einstein, Albert, 12
Eisenhower, Dwight D., 11, 24, 28, 109
Ellsworth, Robert F., 47
El Salvador, 55
Employment Research Associates, 31-32
environmental destruction, 8
Epstein, Joshua, 43, 45
Etzold, Thomas, 22
Europe
   arms control and, 18, 61
   conventional force reductions in
      promote stability, 15, 41, 45, 51, 72, 73, 91
         con, 22, 48, 66
      START and, 78
   international trade, 38
   NATO and, 22-23, 55, 71
   nuclear weapons in, 7, 59, 60
   Soviet Union threatens, 55, 73
      con, 21-22, 45, 51, 95
   START and, 70
   U.S. and, 24-26, 52, 57, 74, 90
European community, 21, 23, 39

France
   and German reunification, 23
   and Great Britain, 68
   in World War II, 64
   nuclear forces, 3, 27, 56
Frank, Barney, 55
Frost, Robert, 11

Gaddis, John, 22
Galvin, John, 54
Gayler, Noel A., 71
Genscher, Hans-Dietrich, 12
Gephardt, Richard A., 34
German reunification, 21, 23, 26
Germany, Nazi, 98
glasnost, 16, 30, 31, 61, 90, 91, 97
Goldman, Marshall, 42
Gorbachev, Mikhail S., 15, 19, 29, 48, 77
   and arms control, 18, 49, 60, 69, 90, 91
   and deterrence, 26
   and eliminating nuclear weapons, 11, 47, 72
   and START, 73

conventional force reductions, 59, 61, 63, 67, 78
reforms of
   U.S. should encourage, 15, 16, 31, 65
      con, 30, 32, 41, 42
Gorshkov, Sergei G., 78
Gottlieb, Alan, 64
Gramm-Rudman-Hollings legislation, 37, 44
Great Britain
   and arms control, 66
   and France, 68
   and German reunification, 23
   economy, 38
   in World War II, 64
   nuclear forces, 3, 27, 56
   U.S. troops in, 74
greenhouse effect, 8
Grenada, 57
Gromyko, Andrei, 61

Haass, Richard, 60
Harmel, Pierre, 91
Hensley, David, 34
Hinds, Bruce, 106
Hiroshima, 2, 4, 11
Hitler, Adolf, 2, 23, 28, 64
Hobson, Art, 101
Howard, Michael, 17
Hungary
   Soviet invasion of, 24
   Soviet troops in, 56, 74
Hunt, Albert, 45
Hutchins, Robert, 4

Iklé, Fred Charles, 21
Incidents at Sea Agreement (1972), 85
India, 68
Inman, Bobby, 80
intercontinental ballistic missiles (ICBM), 102, 109
   arms control and, 16-17, 70
   Soviet, 29-30, 110, 111
   Stealth bomber and, 99
   U.S., 49, 53, 61
Intermediate-Range Nuclear Forces (INF)
   Treaty (1987), 30, 69, 77
   and deterrence, 3
   is beneficial, 59, 60, 61
      con, 63, 64
   is verifiable, 91
      con, 11-12
Iran, 57, 86
Iran-Iraq war, 13, 48, 57
Iraq, 18
   chemical weapons, 57, 86
Israel, 56, 57, 61
Italy, 74

Japan, 26, 33
   defense spending, 8, 31, 37, 72
   economy, 38
   infrastructure spending, 45
   Soviet Union and, 8
   U.S. forces in, 53
Jews, 64
Johnson, Lyndon B., 3, 24
Johnson, Samuel, 23
Jones, John Paul, 83
Jung, Carl, 64

Karber, Phillip, 52
Kasich, John, 41
Kaufmann, William W., 32, 53, 56

Kelly, James, 106
Kennan, George F., 13
Kennedy, John F., 66
Kennedy, Paul, 35, 41
Keynes, John Maynard, 8
Kim Il Sung, 21
Kissinger, Henry A., 71
Korb, Lawrence, 53
Korean War, 22
   military spending, 36
   use of bombers, 102
Kosminsky, Jay P., 73
Krasnoyarsk radar, 19, 29
Kristol, Irving, 59
Kucharski, John M., 34

labor, 37
   medical insurance for, 42
   military spending and, 31-32, 34
Laird, Melvin R., 71
La Rocque, Gene R., 27
Larson, Charles, 79
Latin America, 94
Lehman, John, 53, 57
Lenin, Vladimir I., 91
Libya, 53
   weapons spending, 64
   weapons technologies, 12, 18
Limited Test Ban Treaty (1963), 66
Litvinoff, Maxim, 63

McIver, Robert, 93
McNamara, Robert S., 3, 52, 69
Madrid Mandate, 85
Mandel, Michael J., 31
Marshall, Burton, 67
Melman, Seymour, 34
Milken, Michael, 43
Millett, Allan, 58
Mitchell, George J., 34
Mitterand, François, 41
Moynihan, Daniel Patrick, 44
Muste, A. J., 93
Mustin, Henry, 78
Mutual and Balanced Force Reductions, 91

Nagasaki, 4, 11, 102
National Guard, 45
national security
   arms reduction and, 7
   defense spending and, 27, 55-56
   maritime strategy to protect, 84
Natural Resources Defense Council, 79, 86
New Zealand, 81
Nietzsche, Friedrich, 66
Nitze, Paul, 80
Non-Proliferation Treaty, 13, 61
North Atlantic Treaty Organization (NATO), 51, 55, 61
   and arms control, 59, 90, 91
   and Eastern European revolutions, 21, 22, 23, 31
   is irrelevant, 28
      con, 84
   nuclear strategy, 3, 30, 70, 71-72
   troop reductions, 45, 56-57, 67, 73, 74, 89
   weapons spending, 64, 90
North Korea, 86
   nuclear program, 18
Northrop Aircraft, 105, 106, 107, 108, 109, 111-112
nuclear-freeze movement, 93-94
nuclear war, 1-2, 11, 15, 26, 111

accidental, 42
arms control and, 17, 64, 66, 69
at sea, 78
can be fought, 3, 4
    con, 4-5, 7-8, 9, 71
threatens U.S., 48, 63
    con, 41
nuclear weapons, 97
and defense spending, 49
and deterrence, 48, 111
deaths from, 64
defense systems
    are possible, 3
    con, 5, 70-71
elimination of
    is possible, 11
    con, 12, 56, 72, 94
MX missile, 101
NATO, 59, 71
Soviet, 23, 47
strategic, 29, 48, 52, 55, 70
submarine, 79, 80
tactical, 3, 77, 81
testing ban on, 13
theater, 30
Third World, 27-28
U.S., 27, 48
nuclear winter, 2
Nunn, Sam, 17, 80, 109
Nye, Joseph S., Jr., 59

Ogarkov, N.V., 47-48

Packard Commission, 50
Pakistan, 61
Panama, 52, 53
Partial Test Ban Treaty, 13
"peace dividend," 32
    see also defense spending, U.S.
peace movement, 2, 5, 81, 93-94
Pennar, Karen, 31
perestroika, 15, 16, 30, 31, 90
Perle, Richard, 52
Persian Gulf, 24-25
    oil supplies, 57
    U.S. military operations in, 53, 80, 102
Peter I (emperor of Russia), 83
Philippines, 57
Phillips, Howard, 59
Poland
    Soviet troops in, 51, 56, 74
    U.S. aid to, 43-44
Powell, Colin, 51, 57

Reagan, Ronald, 28, 63, 65, 70
and Afghan resistance, 24
and arms control, 59, 66, 79, 81
and nuclear war, 11, 71
and Soviet Union, 60-61
and Star Wars, 3, 44, 70
arms buildup, 31, 34, 36, 53, 61
social program spending, 42-43, 47
Record, Jeffrey, 101
Reykjavik summit, 11
Rhinelander, John, 56
Richardson, Elliot L., 71
Rittenhouse, John D., 32
Rockwell International Corporation, 33
Rogers, Herbert F., 33
Ross, Michael L., 77
Ross, William S., 33
Russell, Bertrand, 2
Rust, Matthias, 43

Sachs, Jeffrey, 44
Samuelson, Paul A., 31
satellites, intelligence
    and arms control, 60, 90
    and Stealth bomber, 110
    U.S., 54, 57
Saudi Arabia, 57
Schell, Jonathan, 2, 4, 13
Schelling, Thomas, 59-60
Schlesinger, James R., 56, 71
Schmidt, Helmut, 71
Scott, Thomas, 55
Scott, William B., 105
Scowcroft, Brent, 66, 74
Scowcroft Commission, 70
Shevardnadze, Eduard A., 26, 73-74, 75, 79
Shultz, George P., 44
social programs
    should be funded by defense cuts, 41-46
    con, 47-50
South Africa, 61
South Korea
    defense spending, 37-38
    U.S. forces in, 46, 53
Soviet Academy of Sciences, 79, 86
Soviet Union, 93, 94
    and deterrence, 2, 3
    arms control with
        INF, 12, 69
        is beneficial, 5, 7, 60-61, 62
        con, 64
        START, 44, 74
        is necessary, 17, 18-19, 69
    conventional force reductions, 32, 49, 57, 63, 74, 90
    defense spending, 9, 39, 67, 90, 111
    economy, 8, 15, 31, 38, 41, 42, 55, 90
    European forces, 24, 32, 66, 73, 75
        are a threat, 18, 25, 48, 56, 66
        con, 22, 45, 51
    interventions, 24
    navy, 77, 78-79, 80-81, 83, 84, 85-86
    nuclear forces, 53
        threaten U.S., 23, 27, 29-30, 41-42, 47, 48, 49, 56
        con, 43, 100
    reforms, 28
        U.S. is adapting to, 15-19
        con, 21-26
    strategic defense, 70-71, 97-98, 111
    submarine forces, 69-70
    U.S. relations, 18
        are greatly improved, 8-9, 16, 59
        con, 26, 65, 66, 67
    U.S. defense strategy and, 26, 51-52, 56
    Star Wars, 44, 73-74
    Stealth bomber, 99, 101, 105, 110, 111
Stalin, Joseph, 16, 22, 23, 25, 26
Stealth bomber, 48
    cost of, 52-53, 58, 97, 101, 102, 109, 112
    is necessary, 42, 97-100
        con, 101-103
    should be funded, 105-108
        con, 32, 41, 46, 109-112
    START and, 74
Steinhoff, Johannes, 71
Stockholm Accord (1986), 85
Strategic Air Command, 26, 97, 98, 103
Strategic Arms Limitation Talks (SALT), 60, 66, 77, 93
Strategic Arms Reduction Talks (START), 30, 48, 59, 77
    bombers and, 99, 110

naval forces and, 78, 79
promote U.S. interests, 19, 41, 53, 69-72
    con, 16-17, 65, 66, 73, 75
Soviet Union and, 17, 44, 78
U.S. and, 55, 56, 62, 90
verification problems, 65
Strategic Defense Initiative (SDI), 3, 5, 61, 71
cost of, 44, 52-53, 54
research, 48, 67
    should be cut back, 54
    con, 30
should be built, 68
    con, 28, 44, 45, 46, 70
START negotiations and, 17, 70, 73-74
Strong, Robert A., 1
submarine-launched missiles, 17, 19, 77, 81
START reductions, 44, 56, 74, 78, 79
    are beneficial, 69-70
    con, 79-80
U.S., 69, 85, 109
    as deterrent, 42, 43, 53
    should expand use of, 30, 49
    con, 27, 53-54, 80
Sun Tzu, 64
Swomley, John M., 93
Syria, 18
    chemical weapons, 57
    weapons spending, 64
Szilard, Leo, 4

technology
    nuclear defense, 3, 5
    weapons, 12, 16
terrorists, 52, 55, 87
Thatcher, Margaret, 8, 69
Third World, 62, 72
    debt crisis, 44
    nuclear capacity, 27-28
    U.S. policy toward, 16, 25, 57, 78, 80, 102
    weapons technologies in, 18, 64
Thompson, E.P., 2
Thucydides, 1, 18
Trost, Carlisle A.H., 79, 83
Tsipis, Kosta, 7
Turkey, 74
United Nations, 4, 18, 93
    Charter, 62
    peacekeeping forces, 62
United States
    Air Force, 52, 97, 102, 109
        spending should be cut, 54
        Stealth bomber will benefit, 105, 106, 108
            con, 41, 98-99, 101, 103, 110, 112
    and arms control, 7
        benefits of, 12, 17, 23, 52, 53, 61-62, 69
        harms of, 66, 93, 94
        should negotiate START, 69-72
        should reduce conventional arms, 73-75
        should reduce naval weapons, 77-78
            con, 83-87
    Army, 53, 54, 55, 57
    budget deficit, 27, 30, 31, 32, 36-37, 46, 55
    Central Intelligence Agency (CIA), 42, 43, 57, 93-94, 95
    Congress, 51, 95
        and arms control, 17
        and drugs, 94
        and foreign aid, 44
        and nuclear weapons, 28, 49, 78, 101
        and Star Wars, 44, 71
        and Stealth bomber, 41, 99, 103, 108, 109, 112

defense spending, 21, 34, 52
  should cut, 27, 32, 48, 50, 53
    con, 26, 36, 55, 57
Congressional Budget Office, 34
Defense Department, 27, 32, 36, 110
  and arms control, 44-45, 75
  and base closings, 50
  and budget cuts, 30, 31, 37, 52, 53, 54
  and Stealth bomber, 97, 99, 100, 109, 112
  and Third World, 57
  overestimates Soviet threat, 21, 22, 24, 25, 26
    con, 79
European forces, 24, 25, 57
  should be reduced, 22-23, 32, 73-75
Federal Reserve Bank, 33
foreign policy
  toward Eastern Europe, 23-24, 25, 43
  toward Third World, 16, 25, 44, 57, 78, 80, 102
Joint Chiefs of Staff, 22, 27, 70
Marine Corps, 53
Navy, 52, 110
  aircraft carriers, 53, 85, 86
  arms cutbacks, 55
    should be made, 53, 54, 77, 80, 81
      con, 83, 85, 86
  opposition to arms control, 44, 79, 80
nuclear strategy, 48, 49
  must change, 21-26, 101-102
    con, 15-19, 55, 56, 58, 70
  strategic defense, 70-71
Soviet relations, 18, 59
  have greatly improved, 8-9, 16
    con, 65, 66, 67
  must change, 21-26
    con, 15-19
  warrant defense cuts, 51-54
    con, 55-58
State Department, 75
trade deficit, 27, 46
Transportation Department, 45
Treasury Department, 37, 44
USS *Coral Sea*, 86
USS *Nimitz*, 86
USS *Stark*, 57, 80
USS *Vincennes*, 80

Vietnam War, 22, 25, 48
  bombers in, 102
  military spending during, 34, 36

Walcott, John, 55
Warner, John, 80
Warnke, Paul, 66
Warsaw Pact, 23, 26, 53, 84
  is a threat, 55, 89, 91
    con, 21, 22, 24, 28, 31, 51
  nuclear forces, 71
  troop reductions, 45, 56
Webster, William, 53
Weidenbaum, Murray, 35
Weinberger, Caspar, 35, 36, 70, 112
Welch, Larry D., 99, 110
Western Europe. *See* Europe
West Germany, 33
  and arms control, 64, 66
  and NATO, 56-57
  defense spending, 8
  terrorism in, 55
  U.S. forces in, 24, 26
Wolfowitz, Paul, 57
Woolsey, R. James, 43, 66

World War II, 11, 38
  bombers in, 97, 98, 102
  chemical weapons in, 64
  military spending during, 36

Yazov, Dmitri T., 18